High Resolution Site Surveys

High Resolution Site Surveys

Roger Parkinson

CRC Press
Taylor & Francis Group
Boca Raton London New York

CRC Press is an imprint of the
Taylor & Francis Group, an **informa** business

A SPON PRESS BOOK

CRC Press
Taylor & Francis Group
6000 Broken Sound Parkway NW, Suite 300
Boca Raton, FL 33487-2742

First issued in paperback 2019

© 2001 Roger Parkinson
CRC Press is an imprint of Taylor & Francis Group, an Informa business

Typeset in 10/12 Garamond 3B2.603d [Advent] by
Keyword Publishing Services Ltd

No claim to original U.S. Government works

ISBN-13: 978-0-415-24407-7 (hbk)
ISBN-13: 978-0-367-86636-5 (pbk)

This book contains information obtained from authentic and highly regarded sources. Reasonable efforts have been made to publish reliable data and information, but the author and publisher cannot assume responsibility for the validity of all materials or the consequences of their use. The authors and publishers have attempted to trace the copyright holders of all material reproduced in this publication and apologize to copyright holders if permission to publish in this form has not been obtained. If any copyright material has not been acknowledged please write and let us know so we may rectify in any future reprint.

Except as permitted under U.S. Copyright Law, no part of this book may be reprinted, reproduced, transmitted, or utilized in any form by any electronic, mechanical, or other means, now known or hereafter invented, including photocopying, microfilming, and recording, or in any information storage or retrieval system, without written permission from the publishers.

Trademark Notice: Product or corporate names may be trademarks or registered trademarks, and are used only for identification and explanation without intent to infringe.

British Library Cataloguing in Publication Data
A catalogue record for this book is available
from the British Library

Library of Congress Cataloging in Publication Data
Parkinson, Roger, 1946–
   High resolution site surveys/Roger Parkinson
      p. cm.
   1. Seismic prospecting. I. Title.

TN269.8.P37 2001
622′.1592–dc21   00-059527

Visit the Taylor & Francis Web site at
http://www.taylorandfrancis.com

and the CRC Press Web site at
http://www.crcpress.com

To my parents, long gone but not forgotten

Contents

Preface xi
Acknowledgements xiii

1 Review of the site survey technique 1

   1.1 Introduction 1
   1.2 Historical background 1
   1.3 High resolution seismic site surveys 3
   1.4 Digital seismic site surveys 5
   1.5 Analogue site surveys 8
   1.6 Sonar swath bathymetry 10
   1.7 Autonomous underwater vehicles 10
   1.8 Remote-operated vehicles 11
   1.9 Environmental site survey techniques 11
   1.10 Gravity and magnetics 11
   1.11 Seismic refraction 12
   1.12 Pipeline surveys 13
   1.13 Marine archaeological surveys 13
   1.14 Positioning systems 14
   1.15 Site survey grids 15
   1.16 Project planning and preliminaries 16
   1.17 Survey vessels 17
   1.18 Reporting 21

2 High resolution digital site survey systems 24

   2.1 Digital recording systems 24
   2.2 Texas Instruments digital field system (DFS) Mk V 28
   2.3 Sercel SN 358 digital recording system 30
   2.4 TTS-2 31
   2.5 OYO DAS-1A 34

viii   *Contents*

    2.6   *Geometrics Strativisor NX seismic recorder 36*
    2.7   *Digital systems tests 37*
    2.8   *Sparker seismic energy sources 43*
    2.9   *High frequency seismic airgun sources 50*
    2.10 *Other site survey seismic energy sources 57*
    2.11 *Source fire control systems for site surveys 58*
    2.12 *Streamers for site surveys 60*
    2.13 *Onboard processing 66*
    2.14 *Digital site survey interpretation 73*
    2.15 *Contractual work standards 78*

3 Analogue site survey systems   81

    3.1   *Categories of analogue systems 81*
    3.2   *Echosounders 81*
    3.3   *Sidescan sonars 85*
    3.4   *Sonar swath bathymetry 95*
    3.5   *Autonomous underwater vehicles 104*
    3.6   *Sub-bottom profilers 109*
    3.7   *Boomers and sparker profilers 115*
    3.8   *Multi-electrode sparkers 123*
    3.9   *Analogue recorders and filters 130*
    3.10 *Seismic refraction survey 131*
    3.11 *Site survey discussion 132*

4 Non-seismic site survey techniques   134

    4.1   *Gravity corers and seabed sampling 134*
    4.2   *Shearvane penetrometers 135*
    4.3   *Vibrocorers 136*
    4.4   *Underwater photography 137*
    4.5   *Current meters 137*
    4.6   *Engineering tests and cone penetration 138*
    4.7   *Remote-operated vehicles 138*
    4.8   *Gravity survey 141*
    4.9   *Marine proton magnetometers 145*
    4.10 *Marine caesium vapour magnetometers 148*

5 Positioning systems   154

    5.1   *Introducton 154*
    5.2   *Principles of GPS operation 157*
    5.3   *Co-ordinate systems 160*

Contents ix

    5.4   *Geoids, ellipsoids, datums and shifts 161*
    5.5   *GPS differential mode principles 165*
    5.6   *Real-time kinematic (RTK) GPS 166*
    5.7   *DGPS errors 169*
    5.8   *Skyfix, Veripos and Starfix 172*
    5.9   *DGPS quality control 175*
    5.10 *Supervisory quality control systems 177*
    5.11 *Integrated navigation systems 181*
    5.12 *Acoustic positioning systems 186*

6 Safety   192

    6.1   *Safety reviews 192*
    6.2   *Documentation 194*
    6.3   *Health and safety policy 195*
    6.4   *Incident and accident reporting policy 196*
    6.5   *Cranes and lifting gear 196*
    6.6   *Working conditions and working practices 196*
    6.7   *Fire protection equipment 201*
    6.8   *Vessel safety and survival equipment 201*
    6.9   *Training and emergency response procedures 202*
    6.10 *Seismic operations 204*
    6.11 *Electrical equipment and emergency power supplies 209*
    6.12 *Helicopter operations 210*
    6.13 *Personnel competence assurance 211*
    6.14 *Environmental monitoring 212*

Glossary and notation   214
Index   227

Preface

The preparation of this book has involved the collection of more than 300 references from old survey reports, contractors' data sheets, published and unpublished academic papers that relate to site survey work, old conference reports and papers and references from other published works.

There were four reasons for writing this book. First, the boom-and-bust cycle of oil exploration has ensured that the generations are short in the seismic industry. Few people stay in the industry longer than 5–7 years and the majority leave in the next 'bust' cycle. This author has seen the same stupid and sometimes dangerous mistakes repeated over and over again. Each generation leaving the industry take with them a valuable cache of accumulated practical experience that is not passed onto the next generation. This book attempts to rectify this, since it has been written by a consulting engineer with 25 years in the business.

The seismic industry has used a large range of instrumentation over the years which has been accompanied by often unintelligible handbooks and instruction manuals. The author has, over the years, accumulated many such manuals, system descriptions and hand-out documents, which when deciphered reveal an untapped wealth of knowledge about the seismic and site survey world. This book relies heavily on an extensive reading of such manuals, allied to the practical experience of site survey work.

A third reason for writing this book concerns the status of site surveys within the oil industry. Too often site surveys fall between several stools. Sometimes they are managed under the aegis of drilling departments, who do not have any real understanding of high resolution marine geophysics. If site surveys are the provenance of the exploration departments, the geophysicists are often used to 2D and 3D surveys but have little idea of the intricacies of high resolution work.

The fourth and reason final for writing this book it that it is 23 years since a UK-based geophysicist attempted a textbook dealing exclusively with high resolution site surveys. Site surveys are long overdue for re-evaluation.

The format of the book starts with an informal review of site survey techniques (Chapter 1) and moves to detailed explanations of digital survey systems (Chapter 2), analogue survey systems (Chapter 3), non-seismic survey systems (Chapter 4), positioning systems (Chapter 5) and safety considerations (Chapter 6).

No attempt has been made to denigrate particular equipment or contractors, only to compare and contrast the systems in use. Systems described are included because they are typical, not because they are better or worse than other systems available. This book is an honest attempt by a long-serving consultant engineer to provide an overview of the site survey world.

Roger Parkinson

Acknowledgements

In the preparation of this book the author has ruthlessly exploited his every contact in the oil industry. Particular thanks must go to the three managing directors who started their careers in the seismic business with the author, when we all worked for Fairfield-Aquatronics. These are Jim Sommerville (Fugro-Geoteam, Aberdeen), Trevor Smith (TTS Systems) and Bruce Allen (Exploration Electronics). Their assistance included pointing the author in the direction of other contacts and providing a great deal of technical detail, particularly regarding the digital recording systems currently in use.

Thanks must also be extended to the numerous people who were called and asked for information in the most direct terms possible. There were no brush-offs, only interest that anyone had the temerity to embark on a comprehensive review of the site survey industry. On the acoustics side assistance included Kongsberg-Simrad, Sonardyne, Geoacoustics and the International Marine Contractors Association. Geometrics contributed an explanation of their new digital recording system and their caesium vapour magnetometer. Dave Waters (Independent Consultant) and Ed Dunston (Fugro-Geoteam, Swindon) contributed heavily on the safety side.

A number of published textbooks are quoted here and these have been the subject of permissions to quote as indicated in the text. Some of the description of GPS operation is a precis from *GPS Satellite Surveying* by Alfred Leick, by kind permission of John Wiley and Sons, Inc. The descriptions of actual DGPS systems were provided by Racal, Fugro-Geoteam and Dassault-Sercel who all must be thanked for their co-operation. Milton B. Dobrin's *Introduction to Geophysical Prospecting* has been used for a description of bright spots with records reproduced by kind permission of McGraw-Hill Publishing Company.

Contract and system specifications for surveys rely heavily on the *UK Offshore Operators' Conduct of Mobile Drilling Rig Site Surveys*, Volumes 1 and 2.

Finally, thanks must go to Geophysical Consultants Ltd (Tony Lucas and Malcolm Wood) who were endlessly tolerant of the requests for information and Bob Barton (Barton Marketing Services), who suggested the project, having seen an earlier handbook written by the author. Thanks are also extended to the four surveyors who read the positioning chapter and passed a long series of useful comment and critique back to the author: Ed Danson, Steve Barmforth, Dave Bee and Martin Joseph.

1 Review of the site survey technique

1.1 Introduction

From time to time most oil companies conduct high resolution and engineering site surveys. It is often the case that these site surveys are conducted at short notice for reasons such as rig insurance, drilling department requirements, partnership arrangements, geophysical problems or government regulation. Site surveys delineate possible obstacles or dangers to drilling, pipeline laying or the erection of structures on the sea floor. Many drilling hazards are now widely recognised and these include buried channels, weak strata, shallow gasification and shallow faulting. These potential dangers have varying importance for different types of structure and surveys must be designed around the type of structure being placed at a particular location. This book attempts to analyse the site survey problem as experienced by oil companies. The process of contractor selection is discussed and a detailed analysis of data acquisition and quality control is made.

High resolution survey techniques are also used widely outside the direct confines of seismic and site survey requirements for projects that include marine archaeology, pipeline investigation, marine cable laying, mineral exploration/exploitation, civil engineering and university-based geological and geophysical projects.

Most oil companies or other enterprises who conduct site surveys appoint a consulting engineer to help set up and supervise a proposed survey programme. This book has been written from the perspective of a long-serving consultant engineer and views the site survey world from as wide a viewpoint as possible.

1.2 Historical background

The general theory of seismic waves can be dated as far back as Hooke's law of 1678. For many materials subjected to small applied forces such as seismic waves, deformation is proportional to the applied force. By the early nineteenth century compressional or P-waves and shear or S-waves were differentiated. P-waves are longitudinal waves where the direction of particle motion is parallel to the direction of wave propagation. In S-waves these

2 *High resolution site surveys*

directions are perpendicular to each other. During World War I, mechanical seismographs were used to locate enemy artillery, while early experimental hydrophones were used to position ships. The work of Langevin and Chilowsky led to the piezoelectric oscillator and this, harnessed to a three-electrode (triode) valve amplifier, produced a true sonar system for submarine detection.

Box 1.1
A historical digression
During World War I, early and very primitive acoustic techniques were used to position bombardment ships off the Belgian coast. The bombardment ship, known as a monitor, had to fix its firing position in relation to the target location even when out of sight of land. A buoy would be laid in advance at the designated firing point and two depth charges were detonated, one on each side of the buoy. Hydrophones situated off the Kent coast of England received the sound of the explosions and after cross-plotting the position signalled the exact buoy position to the monitor.[1] Knowing the exact position of the ship allowed long range guns to be directed onto enemy targets as much as 20 miles away.

In the 1920s the idea of seismic refraction as a tool for oil exploration was becoming accepted. By 1929, refraction seismic surveys had located some fifty salt dome structures. Seismic reflection surveys followed hard on the heels of seismic refraction surveys. In marine seismic reflection surveys P-waves are used. Marine seismic sound sources generate P-waves almost exclusively. This will always be the case in water because fluids have a zero resistance to shear and cannot sustain S-waves. It should be understood that P-waves can generate S-waves when a P-wave strikes strata interfaces at inclined angles of incidence.[2]

In the inter-war period a good deal of basic research was done in Britain and America on magnetostrictive devices. Most modern pingers and other high resolution profiling devices are descended from the early magnetostrictive devices of this period. In Britain an effective strip chart recorder was developed. In terms of basic physics it was realised that underwater sound waves behaved similarly to light waves and were subject to the same laws of reflection and refraction, while the sea acted as a variable refractive medium.[3]

World War II and the development of submarine detection devices ensured that acoustics was a well understood science in the post-war world. The Rayleigh–Willis curve allowed a direct comparison of all acoustic energy sources on the basis of energy or waveform shape. As far back as 1917, Rayleigh related bubble frequency to bubble radius, pressure and fluid density. In the Rayleigh equation,

$T = 1.83 A_m \sqrt{\rho/p_0}$

where T is the period of bubble oscillation (s), A_m is the maximum radius of bubble, ρ is the specific gravity of fluid (Gm/cc) and p_0 is the ambient absolute hydrostatic pressure (dyn/cm³). In 1941, Willis expressed this relationship in terms of source energy (Q in joules),

$$Q = \tfrac{4}{3}\pi A_m^3 p_0$$

where Q is the potential energy of bubble and A is the radius of bubble. The oscillation period T for the bubble effect varies as the cube of the energy Q (in joules) and inversely as the 5/6 power of the pressure. Generally speaking, large energy outputs produce low frequencies:[4]

$$T = 0.0450 Q^{1/3}/(D+33)^{5/6}$$

where D is the depth in feet.

Magnetic tape recording became possible in the early 1950s and this made common depth point (CDP) recording a practical proposition. CDP recording gives a huge increase in the signal-to-noise ratio of the received seismic signals by stacking a number of seismic impulses to a single 'common' depth point. This principle is illustrated in Figure 1.1. At the end of the 1960s, digital recording revolutionised the seismic industry. Non-explosive seismic sources, the common depth point method and the digitisation of records made very sophisticated signal processing techniques possible. Seismic surveys as we know them today came into existence as a result of these three elements.

1.3 High resolution seismic site surveys

Between 1955 and 1972 there was a major oil-rig blow-out approximately once a year. The central problem confronting the oil industry was the necessity to accurately delineate drilling problems, particularly shallow gas accumulations, in the first 1000 m of strata below the seabed. This was the depth to which the oil-rig 30-inch casing was laid and the well spudded with a blow-out protector in place. 2D survey systems were virtually useless for shallow data because the sound source frequency (6–40 Hz) was too low for high definition surveys in shallow strata. High frequency (400–3500 Hz) single channel profiling systems were widely used, but below depths equivalent to the depth of water it proved impossible to accurately interpret the seismic data due to the presence of 'multiples'. At this time nearly all exploration wells were drilled in the shallow waters of shelf seas and the problem of well blow-outs appeared intractable. Indeed the first 1000 m of strata beneath the sea floor became known as the 'twilight zone'.[5]

It was then realised that applying the technique of digital seismic processing, normally used for multi-channel 2D data, to high frequency data recorded digitally, solved the problem of multiple removal from the seismic

4 High resolution site surveys

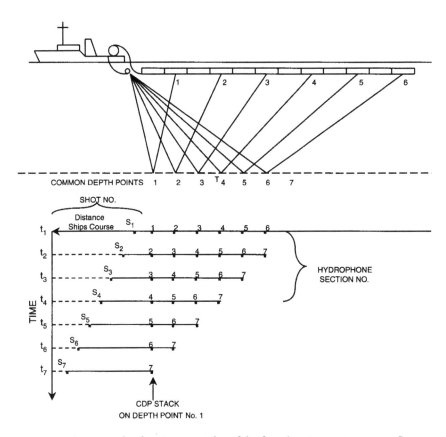

Figure 1.1 Common depth point principles. If the first shot S_1 is at time t_1, a reflection from the depth point number 1 is received on the first streamer section and recorded on board the survey ship as channel number 1. At the next shot S_2 the survey ship will have timed its progress to a position such that the reflection point in section two of the streamer is from the identical depth point position as was received in section 1 from shot S_1. Successive shots $S_3, S_4, S_5, \ldots, S_n$ follow and are stacked to the same common depth point. As the survey ship progresses along its course the seismic signal from CDP 1 travels an ever increasing distance between the shotpoint and the streamer section. If this geometry change is compensated for, all the shotpoints can be stacked for a particular common depth point.

record. The use of a high frequency energy source and 'bright spot' analysis could identify shallow gas accumulations. The 'bright spot' technique is based on the principle that gas-saturated sands have a lower transmission velocity than water or oil-saturated sands. Velocity contrasts across surfaces bounding the gas zones above or below, give reflections of higher amplitude than would be observed from the same interface on either side of the gas zones.

1.4 Digital seismic site surveys

Digital seismic acquisition for site surveys is a miniaturised high resolution version of the 2D seismic survey technique in use for many years. In this technique, referring back to Figure 1.1, a shot at the near trace reflection point is received by the first streamer hydrophone group. If the trace interval is 12.5 m, a second shot 12.5 m after the first will also be summed to this common depth point. Successive shots will then be summed to a single common depth point. It should also be noted that the sub-surface coverage is always half the surface coverage. Figure 1.2 illustrates this point diagrammatically. If there were twenty hydrophones per group and eight shots at the same common depth point, then during processing twenty-four-fold stacking would give an improvement in the signal-to-noise ratio of[6]

$$\sqrt{8 \times 20 \times 24} = 62 \text{ or } 36 \text{ dB}$$

Digital site survey data is used not just for shallow gas (bright spot) analysis but also for shallow strata delineation up to 2–3 s of two-way travel time beneath the seabed. Such data is often an insurance requirement because a rig is unprotected against blow-outs until the 30-inch casing and blow-out protector are in place. It should also be understood that bright spot analysis is the only known seismic technique for delineating drilling hazards such as

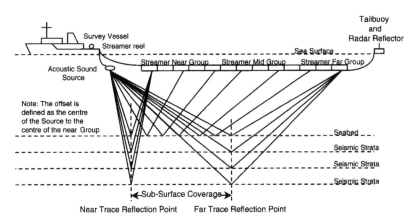

Figure 1.2 Seismic reflection points and sub-surface coverage. This diagram illustrates the fact that the sub-surface coverage is always half the surface coverage. Nearly all site surveys use streamers with a 12.5 m group interval. For full-fold coverage the source firing interval must be half this, that is 6.25 m. Such a short firing interval is often not possible due to the limitations of the digital recording system, compressor limitations at a firing rate of twenty shots per minute and considerations of vessel speed. Full-fold coverage is desirable but often not achievable. This point is sometimes misunderstood within the oil companies.

6 *High resolution site surveys*

Box 1.2
The birth of digital high resolution geophysics
The use of digital seismic survey techniques to locate high pressure gas pockets was one of the principal technical breakthroughs of the seismic industry in the 1970s. The first seismic survey company to use 'bright spot' analysis allied to a high frequency seismic sound source and digital recording was Fairfield-Aquatronics. In the early and mid 1970s they were also the first survey company to record digital data at a sample rate of 1 ms. The author of this book started his career in the seismic industry with this company.

---

shallow gas. A realistic assessment of the survey capability of most contractors suggests that a gas accumulation of 5 m thickness and 200 m diameter should be detectable at 1000 m below the seabed.[7] It should also be clearly understood that no survey, however sophisticated the survey systems in use, can guarantee that all gas accumulations will be identified. There will always be some limitation in resolution and data interpretation and this means that site surveys, though useful for delineating shallow gas, are not infallible.

The effectiveness of high resolution seismic surveys depends on a number of factors. These include seismic pulse penetration, vertical and horizontal accuracy and the resolving capacity of the system in use. Penetration is determined by the power and frequency of the seismic source, allied to conditions such as weather, seismic interference, length of streamer, number of channels, etc. In terms of penetration, the absorption characteristics of the strata hold particular significance. Typical seismic pulse absorption will be in the region of 0.2 dB per wavelength. If no more than a 20 dB loss in signal strength can be tolerated, then 50 Hz signals will penetrate 2000 m below the seabed, 100 Hz signals will penetrate to 1000 m, 500 Hz signals will penetrate to 180 m and 2000 Hz signals will penetrate to 45 m.[8]

Vertical accuracy is usually determined by examining mis-ties at line intersections. Vertical resolution is controlled by the time duration of the seismic wavelet, which is itself dependent upon the bandwidth of the transmitted pulse. Vertical resolution is also defined as the ability to separate two very closely spaced seismic reflections perhaps as little as 3–5 m apart.

Horizontal accuracy, in extreme cases can be measured in hundreds of metres, but more typical accuracy for site surveys is tens of metres. Provided that feathering angles on towed hydrophone arrays (streamers), are kept within reasonable limits, then positioning errors are unlikely to be significant. The biggest source of horizontal accuracy error usually arises from complex geology and high dips and this can normally be corrected, at least in part, by migration of data during processing, a procedure which also aids horizontal resolution. Other factors which effect horizontal resolution include the characteristics of the transmitted seismic signal, the trace interval, the hydrophone array length, the survey line spacing, feathering (already mentioned), seismic data processing and seismic wave propagation.[9]

Digital recording systems for site surveys date back to the late 1960s. Early systems used first-generation digital computers with a number of dedicated interfaces. These were quickly superseded by industry standard seismic recording systems. Both binary and quaternary floating-point systems were used at fast sampling rates, typically 1 ms. The limitation of such systems was multiplexer switching and in the 1990s sigma–delta recording systems superseded most of the older systems in use. The advantage of sigma–delta systems is the ability to use fast sampling rates allied to very short shotpoint intervals. Effectively each input channel has its own analogue-to-digital converter and this overcomes the limitations of multiplexer-based digital recording. Typically, these systems operate at a 1 ms sampling rate for 12-, 24-, 48- or 96-channel operation. A typical record length would be 2.0 s with the expectation of data to a depth of at least 1500 m below the seabed.

The seismic energy source is usually a spark array or a small airgun array or perhaps a single airgun. Water guns and sleeve guns can also be used. Generally speaking, sparkers have now been superseded by airguns or sleeve guns of one sort or another. This book describes sparkers in some detail as much for reasons of historical completeness as anything else and a full description of airguns is included. These sources produce a high frequency spectrum at 40–130 Hz, compared to the 5–60 Hz of conventional seismic sources. The power output will be low, in the region 5–12 bar m. There are other older seismic sources, such as mini-sleeve exploders or flexichoc, but these are usually considered obsolete today.

The seismic streamers used for site surveys are usually short versions of the streamers used for 2D seismic surveys, though there are purpose-designed mini-streamers. Site survey streamers are usually 300, 600 or 1200 m in length with 12, 24, 48 or 96 traces and a trace separation of 6.25 or 12.5 m. These streamers are towed at a depth of 3–4 m below the sea surface. Typical streamers used for site survey operations include the older Teledyne or Fjord type analogue streamers with transformer coupling to the digital recording system. Digital streamers such as the Geco–Fjord system are used by at least two contractors. Sigma–delta recording systems now offer the prospect of modern digital mini-streamers for site survey work and one contractor has such a system. Ideally the seismic source chosen should be able to operate at a shotpoint interval of 6.25 m for full fold coverage at a streamer trace interval of 12.5 m. To meet such an exacting specification a sigma–delta recording system should be used. This point is illustrated in Figure 1.2.

Throughout this book, 'digital survey' refers to multi-channel streamers allied to digital recording systems and a seismic energy source of 5–15 bar m output, with a frequency spectrum of 40–140 Hz. On most site surveys the

8 *High resolution site surveys*

digital acquisition package will be run as a separate entity from the analogue systems described below.

1.5 Analogue site surveys

Analogue survey systems are usually single or dual channel instruments, almost always run as an integrated package of two, three or four instruments separate from the digital survey systems. This prevents interference between the different analogue and digital survey systems. For the purposes of this book, analogue site surveys are those that use single or dual channel recording systems as opposed to multi-channel recording devices.

Analogue systems are used to map the seabed and the near surface seismic strata. Until fairly recently, analogue data rarely went below the first multiple and was in consequence limited to a penetration equivalent to the water depth. Digital seismic processing of single channel profiling data can now remove multiples and greatly extend the range of such data. Analogue systems accurately locate seabed obstacles such as wreckage, pipelines and telephone cables and can outline some geological features. Pock marks, surface sediments, outcroppings, bedrock, shallow channels and other features may be mapped with analogue equipment. As a general rule the analogue systems should give at least 30 m of penetration below the sea floor for jack-up rig leg penetration assessment and at least 10 m of penetration for anchor holding prediction. There should also be an overlap between the analogue profiling systems and the digital seismic data.

There are five main categories of instrumentation used for analogue site surveys. The first category of analogue system is the hydrographic echosounder. Echosounders for site surveys should be dual channel hydrographic echosounders, properly calibrated and able to delineate sea depth to within ±10 cm. The resolution of echosounders is determined by beam width and pulse repetition rate. A properly installed echosounder should also be operated with the digital survey package.

The second category of analogue system is the sidescan sonar which gives a picture of the sea floor and delineates obstructions on the sea floor. Each traverse of the survey grid covers an area perhaps 200 m to either side of the survey line in question. Sidescan sonars usually operate at a frequency of 100 kHz and can be used alongside pingers and boomers (described in the following paragraphs). Modern systems digitise the sonar data and record the sonar data to tape. Some systems integrate a profiler transducer with a dual channel sidescan sonar to produce an integrated sonar/profiler record. A recent development is the use of dual frequency sidescan sonars, usually at 100 kHz and 400 kHz.

The third category of analogue instrumentation is pingers and profilers and these are the highest frequency profilers in service. These systems use single frequency outputs usually somewhere between 3.5 and 10.0 kHz; 3.5 kHz is extremely common. Such systems use electromagnetic transducers as a point source of seismic energy. The high frequency data obtained is used to delineate strata immediately below the seabed, to a depth of 5–30 m. Such data is essential for assessing the leg penetration of jack-up oil-rigs and for assessing anchor conditions for moored platforms. Some systems transmit and receive data with a single transducer exactly as an echosounder transmits and receives data. The alternative is to receive data using a single point hydrophone usually contained within the pinger or profiler tow vehicle. Most pingers and profilers are fish-mounted sub-tow systems, positioned at 3–5 m below the sea surface.

The fourth category of analogue system are the deep-tow and surface-tow boomers and sparkers which use either an electromagnetic transducer or an electric spark discharge (sparker), to obtain data deeper than that obtained from a pinger or profiler. Such data is usually in the region 5–100 m below the seabed. The frequency spectrum obtained from boomers is a true broadband seismic pulse and is lower than that obtained from pingers. The frequency spectrum used will be decided on site, typically 800–5000 Hz. Data reception will usually be on a single point hydrophone or a multi-element single trace mini-streamer with typically, an array of nine hydrophones.

Most profilers, sparkers and boomers have surface-tow, sub-tow or deep-tow variants. Surface tow systems are usually mounted on a catamaran sled and towed behind the survey vessel. In good weather conditions the data obtained can be outstanding but the weather window that allows surface tow operations is small. Deep-tow systems use a submersible, usually free-flooding towfish with the boomer or sparker system placed perhaps as much as 20–25 m below the sea surface. This places the seismic element well below sea surface wave effects and ensures that data can be acquired across a wide range of weather conditions, considered essential for North Sea surveys. Deep-tow data is often considered inherently inferior to surface tow data but will fulfil the requirements of most site surveys. Sub-tow systems, placed perhaps 5 m below the sea surface, can have advantages over surface-tow and deep-tow systems when the tradeoff between operational weather windows and overall data quality is made.

The fifth and final category of analogue systems is multi-electrode sparkers, which operate at a frequency typically from 400–2000 Hz, though this can be varied considerably on site. The energy discharge is always an electric sparker type with a sparker discharge unit that has as many as 144 discharge tips (sometimes referred to as a 'comb' sparker). The data is received on a single trace mini-streamer. Multi-electrode sparkers are unsatisfactory in that they cannot usually be operated as part of a package due to interference effects and must therefore be operated in a stand-alone mode. Multi-electrode sparker data is sometimes necessary to obtain data at 20–400 ms below the sea floor.

Most analogue site surveys use a hydrographic echosounder, a high frequency pinger, a lower frequency boomer or profiler and a sidescan sonar all operated as an integrated package.

1.6 Sonar swath bathymetry

Conventional echosounder-derived bathymetry requires a large number of passes to be made across a small survey area, to construct an accurate bathymetric chart. Such work is usually the province of the Royal Navy Hydrographic Service. Swath bathymetry allows the construction of accurate bathymetric charts from a relatively small number of passes across a survey area.

Sonar swath bathymetry uses either multibeam echosounders or phase measurement scanned sonar systems. Sonar-corrected or echosounder-corrected slant ranges give a digitised soundings chart over a large area of sea floor with relatively few passes of the survey area. Both methods are dealt with in Chapter 3 of this book.

Modern swath bathymetry systems often include a pinger/profiler unit as part of an integrated system that can fulfil the needs and requirements of, typically, a pipeline route survey, in one survey unit.

1.7 Autonomous underwater vehicles

In the 1990s the oil industry moved steadily into deep water oil and gas exploration projects. In 1998 for example, over 200 wells were drilled in water deeper than 500 m. The survey techniques and instruments described in Section 1.5 do not really meet the technical or cost specifications for deep water survey. In a water depth of 1000 m a typical sidescan sonar will have 2500 m of cable deployed. With a 3 km run-out at the end of a survey line, followed by a slow turn and a 3 km run-in to the next line, a survey vessel can spend as much time on line turns as actually surveying. On a survey operation costing about £25,000 per day, deep water survey is costly and time consuming. There are other problems associated with deep water survey. With 2500 m of cable deployed, the positional accuracy of the towfish will only be known to within ±30 m, not accurate enough for modern survey requirements. Additionally, the survey vessel speed when towing such a long cable may be as low as 1 knot.

Within the oil industry there is a developing requirement for autonomous underwater vehicles that can carry out multibeam echosounder, sonar swath bathymetry, sidescan sonar and sub-bottom profiler functions. The autonomous vehicle would be deployed from a survey vessel and its position during survey relayed to the survey vessel by acoustic means. The vehicle would not require lengthy and time-consuming line turns and a reasonably fast data acquisition rate should be possible, compared to conventional towed systems.

It should also be remembered that typical site survey grids (see Section 1.15) are constructed around typical North Sea survey conditions, that is, in water depths of approximately 200 m. A 3×3 km grid represents 9 km^2 of seabed to be surveyed. In 1000 m of water the anchor pattern for a moored rig can cover an area of 50 km^2, so the survey area to be covered is very much greater in very deep water.

1.8 Remote-operated vehicles

Twenty years ago nearly all rig and wellhead inspection work was carried out by divers. Remote-operated vehicles (ROVs) were considered expensive and unreliable. The video cameras gave often blurred and unsatisfactory results. Diving was never safe but in shallow water it was a practical proposition. As the technology improved and the oil industry moved into ever deeper waters, where divers could not go, ROVs were used in increasing numbers. The cameras, the electronics and control systems improved steadily until today there is virtually no commercial deep water diving. In the North Sea alone there are approximately 450–500 ROVs currently in service. The steady improvement in ROVs for deep water work has also ensured that they are used in shallow waters where divers would once have been used.

Pipeline inspection is often carried out using ROVs as further described in Section 1.12 of this chapter.

1.9 Environmental site survey techniques

In addition to the survey systems already described analogue site surveys utilise drop corers and grab buckets for sea floor sampling. Vibro-corers might be used for deeper cores. Underwater photography and underwater video cameras can be used for evaluating obstacles or assessing old wellheads that are being reactivated. Environmental monitoring devices such as current meters, seawater sampling, plankton sampling and pollution sampling devices are all used on occasion. During the 1990s, environmental monitoring became much more important than previously and is likely to increase in the twenty-first century. Government agencies frequently require pre- and post-drilling environmental surveys, to ensure that drilling programmes do not interfere with the flora and fauna of a survey area. Monitoring of the rare cold water corals west of Shetland is an example of this type of necessary environmental survey.

1.10 Gravity and magnetics

Gravity and magnetic techniques are sometimes used, though gravity methods are more usually associated with 2D seismic surveys. Gravity survey fell into some desuetude during the 1980s and early 1990s but has become relatively popular again. Many of the modern low cost 2D operations now include gravity survey as part of a standard package and there is no reason not to include gravity survey as part of a site survey package.

Magnetometers are frequently run as part of a site survey package. Magnetic sensing as a means of prospecting for oil and ore has been known about for 100 years. The inevitable pitfalls associated with magnetometers concern the use of sub-contractors and interference effects between the primary contractor's pingers and profilers and the secondary contractor's magnetometer. The older types of proton magnetometer tended to be unreliable, particularly

12 High resolution site surveys

when newly installed or subjected to air-freight handling. The newer caesium vapour magnetometers are more robust for marine use.

1.11 Seismic refraction

If the seismic source and receiver are separated the reflection angles widen and eventually seismic refractions are obtained. Refracted seismic waves travel along the seismic strata and since the travel path is different from the reflection path refracted waves are an excellent means of obtaining velocity data. The critical angle at which reflections cease and refractions predominate is governed by the laws of classical optics, specifically Snell's law (Figure 1.3). Seismic refraction data requirements are not very common for site surveys but occasionally refraction data is required. Within the seismic industry as a whole such data is usually needed for three reasons. First, to obtain data on layering beyond the depth range of the seismic reflection method. Second, to study deep structures at low cost but in less detail than by reflection methods. Third, to obtain velocity data in the absence of well velocity data. In the context of high resolution engineering site surveys the acquisition of velocity data is only of real use for seismic refraction techniques. The older method of obtaining seismic refraction data was by using sonobuoys as transmitter/receivers. The energy source was usually a sparker or airgun(s). The sonobuoy was free-floating or moored. The sonobuoy consisted of a VHF transmitter, frequency modulated by the seismic data which was transmitted back to the ship. Two time measurements were made, the shot interval to the first arrival and the shot instant to the direct water-wave arrival. The second time divided by the velocity of sound through seawater gave the distance between the shotpoint and the sonobuoy.

One contractor now offers an integrated seismic refraction/reflection system used mostly for pipeline route surveys. This system has a sledge

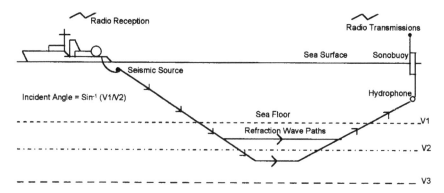

Figure 1.3 Seismic refraction principles. Refraction of seismic waves occurs when the angle of incidence reaches a critical value governed by Snell's law. The angle at which refraction occurs, $I = \sin^{-1}(V1/V2)$.

towed on the sea floor, with a streamer towed behind. A stop-and-go motion device ensures that the sledge is stationary and the streamer in contact with the seabed when the acoustic pulse is initiated. The integration of refraction and reflection data gives very high definition control of the seismic velocities in shallow strata.

1.12 Pipeline surveys

Pipeline route surveys and site investigations are limited to shallow information. Normally a pipeline survey will employ an echosounder, sidescan sonar, pinger/profiler and a gravity corer. The echosounder and sidescan sonar will ensure that the route is free from seabed obstructions and will pinpoint areas of sharp slopes, trenches, etc. The pinger and gravity corer will together form a picture of the geology of the first metre or so of the bottom sediments. Sonar swath bathymetry is a modern alternative method of performing pipeline route surveys.

It is sometimes the case that sidescan sonar and single trace seismic reflection profiling provide continuous data but imprecise information about the characteristics of the very shallow soils, required for the assessment of cable burial. A combination of seismic reflection and seismic refraction can then be used as described in Section 1.11 above.

Pipeline surveys are often carried out for inspection and maintenance purposes on existing pipelines. The pinger, sidescan sonar and echosounder are again the standard instruments. An alternative method of pipeline inspection is an ROV submersible camera unit that is lowered onto the pipeline and crawls along the pipe on rollers, with inspection cameras to port, starboard and above. Pipeline inspection can check on obstructions, pipeline damage or hazard debris. Typical inspection policy is to examine all pipelines with sidescan sonar every year and to perform a detailed ROV survey every 5 years. The main objective with the sidescan sonar is to identify 'spans', that is, areas where scouring has caused the pipeline to become unsupported. At the millennium it appears that about 5000 km of offshore pipeline is laid per annum.

Examples of debris that can become entangled with a pipeline include anchor chains and fishing tackle. This type of debris may in turn foul a submersible used for welding inspection. Low cost surveys that delineate these hazards in advance can save a great deal of time and trouble.

1.13 Marine archaeological surveys

As the oceans and seas of the world have become ever more opened up to the oil industry, so wrecks, some ancient and some relatively new, have been discovered. Most have been discovered inadvertently and have been identified from sidescan sonar records. Occasionally the wrecks and other debris have national historical importance. Greek oared galleys, Roman ships,

14 *High resolution site surveys*

Spanish treasure ships and of course the large numbers of wrecks from the two world wars have all been seen. If government agencies require an archaeological survey to be performed this is usually not a problem. The question is always, who pays for the survey. Most archaeological surveys utilise an echosounder, sidescan sonar and pinger/profiler. Spectacular wreck finds such as the *Titanic* and the *Bismarck* are usually the result of deliberate wreck searches but this need not always be the case. After a wreck or archaeological find has been made with sidescan sonar, further survey work with an ROV or perhaps in future with an AOV may be necessary.

Recent wreck finds include the *Liberty Bell 7* space capsule, launched in 1961. This capsule was recovered 302 miles from Cape Canaveral, Florida at a water depth of 4700 m. The American Civil War wreck *Monitor*, the world's first turret ironclad, was found off Cape Hatteras, using sidescan sonar.

1.14 Positioning systems

A few years ago there were numerous terrestrial positioning systems in use. Names such as Syledis, Hyperfix, Argo and Maxiran were commonplace. One decisive change that took place in the 1990s is the replacement of all terrestrial positioning systems with differential mode global positioning as the means of positioning for virtually all site surveys and 2D/3D seismic surveys.

The Navigation Satellite Timing and Ranging (NAVSTAR) Global Positioning System (GPS) was designed as a second generation satellite positioning system. Development started in the late 1970s and the system became operational in 1982. There are two transmitted codes, known as the precision (P) code and the commercially available (CA) code. For commercial survey work only the second CA code is available and this is not accurate enough for most geophysical survey applications. As a result, differential mode GPS was developed and the positioning accuracy of most surveys is better than ~ 5 m.

The fundamental principle of DGPS is the comparison of the position of a known fixed point, referred to as the reference station, situated on shore, with positions obtained from a GPS receiver at that point. This data is used to calculate corrections to the satellite range data. These differences are then transmitted to the survey vessel which applies the corrections to its own satellite navigation data. Provided the baseline between the reference station and the survey vessel is not too great, the differential corrections improve the quality of the navigation fix to ± 5 m, compared with an uncorrected position that may be ± 100 m. In recent years the navigation signals have been deliberately degraded, to protect the military interests of the United States, for whom the GPS was originally built. This has been overcome by using fast update rates, typically 1 s, transmitted over commercially available satellite channels. Differential mode GPS is likely to be used for many years to come and there is at present no foreseeable successor system. Figure 1.4 shows some of the more commonly used positioning terminology.

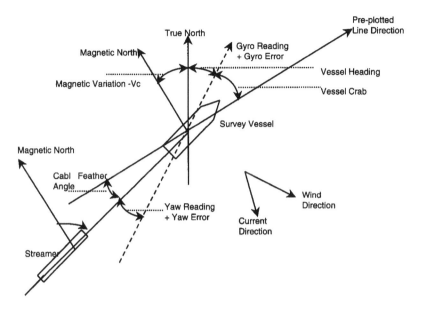

Figure 1.4 Some commonly used positioning survey terms.

1.15 Site survey grids

Site survey grids vary according to the requirements of a particular survey. A typical site survey grid is 3 × 3 km, though grids as small as 2 × 2 km and as large as 10 × 10 km may be used. Line spacing is usually determined by the sidescan sonar and bathymetry coverage requirements. Having determined the line spacing, typically 100 m, the pinger/profiler and boomer would be run alongside the sidescan sonar and echosounder. The entire grid will usually be acquired twice, once with the analogue seismic systems and once with the digital seismic systems. It should be emphasised that there are no hard and fast rules for the construction of survey grids. The digital seismic acquisition grid does not have to mirror the analogue seismic survey grid. Equally there does not have to be exactly the same number of lines in the two grid directions. The construction of a survey grid is very much a factor of the particular survey in hand at a particular moment in time. Figures 1.5 and 1.6 show fairly typical site survey grids with different grids for the digital and analogue systems.

Survey grids are designed to provide 'optimum coverage'. This is a deliberately loose term as 'optimisation' depends on many factors. The grid should be close enough so that problematical zones and structures may be mapped but cost is also a limit on line spacing and on the total number of lines acquired. If the site survey is conducted for a jack-up rig moving onto location then the grid lines are often as close as 50 m in the vicinity of the proposed rig position. A number of tie-lines should also be run with preferably diagonal lines across the proposed wellhead.

16 *High resolution site surveys*

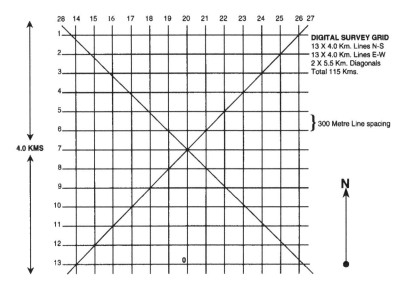

Figure 1.5 Typical digital site survey grid.

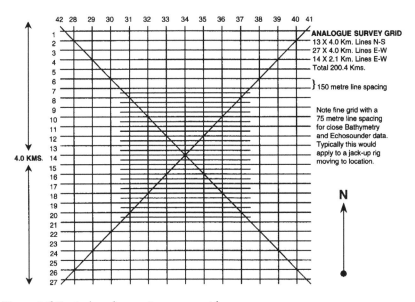

Figure 1.6 Typical analogue site survey grid.

1.16 Project planning and preliminaries

The first stage of setting up a site survey programme is to define the survey objectives in a very precise manner. This is usually done by the oil company geophysicists, though the consultant engineer may be involved at this early stage. The sort of questions that might reasonably be asked at this stage include the following.

- Is there any existing geotechnical data available for the area in question?
- Is there any top-hole well log data that can be used to calibrate the seismic interpretation?
- Are there any shallow gas indications from previous 2D/3D seismic surveys?
- Is there any previous site survey data from the general area of the survey?
- Is there any existing data relating to jack-up rig leg penetration, displacement or scour records for the area in question?
- Is there any existing data relating to anchor holding and guide base stability for the area in question?
- Can old 2D or 3D seismic data be reprocessed for high resolution bright spot analysis?

After writing the specifications for a particular site survey the invitations to bid can be made to the contractors who perform site survey operations. The choice of contractor can then be made. This will depend on a number of factors of which cost is, realistically, the most important. The choice of vessel is also critical and the wrong vessel can ruin a site survey.

When evaluating contract bids for site survey work, particular attention should be paid to the survey vessel proposed by the contractor. The range of vessels used for site surveys varies from purpose-built ships, through to converted vessels permanently fitted with survey gear, to vessels hastily mobilised for a particular project or perhaps a series of projects. A recent innovation is the use of older, single streamer 2D seismic vessels as permanently mobilised site survey operations.

Having considered the above the correct choice of contractor should be made on the basis of a number of interactive factors:

- *Historical experience.* Has this contractor worked for a particular oil company before? What sort of performance was achieved? Was the previous survey completed to cost and satisfaction?
- *Technical prowess.* Can a particular contractor meet the survey parameters and work standards of the survey in question?
- *Vessel considerations.* A survey bid at a low price may be a potential disaster because a cheap but wholly inadequate survey vessel is proposed.
- *Cost.* The lowest bid is not necessarily the best bid when the above three factors are considered.

1.17 Survey vessels

In considering the choice of survey vessel in order of preference, obviously the purpose-built vessel is preferable, though these are few and far between in the site survey market and tend to be expensive. A permanently mobilised operation, utilising perhaps a converted stern trawler or old single streamer seismic vessel, represents the best tradeoff between cost and a reliable seagoing operation. This type of operation is essential for North Sea work where weatherly sea keeping qualities are critical. If a newly mobilised operation is offered it might be preferable to wait until the contractor has completed one or two surveys before commencing the survey in question. There are always problems with newly mobilised operations and contractors are unwilling to charter vessels if they cannot charge the client for the time. Additionally, survey companies will attempt to reduce the mobilisation time if at all possible, by cutting corners on a too-rapid mobilisation. It is almost inevitably the case that newly mobilised operations will be, in some sense, accident prone.

Box 1.3
Cautionary tales of the wrong choice of survey vessel
Choosing the wrong site survey vessel can cost a great deal of money and lost time. One such vessel seen by this author produced so much acoustic noise at any engine revolutions above dead slow that the digital records were virtually useless. Another was used for surveying over a reef structure. Headed up into wind and sea the survey vessel nearly stopped. The Captain was not experienced in site survey work and did not increase speed quickly enough. The seismic streamer grounded on the reef and was so badly holed as to be unusable. The survey was ruined as a result. The case of a purpose-built survey vessel that was acoustically noisy and had to be abandoned is too well known in the seismic industry to need further comment.

Particularly in the case of hastily mobilised site survey operations in areas of the world such as West Africa, the choice of vessel is often a rig supply boat and many unsuccessful site surveys have resulted from this type of operation. A vessel proposed for site survey work should have, in addition to adequate accommodation, a variable-pitch main propeller and a proper autopilot such a Decca Arkas. It should also have a dedicated instrument room power supply. In the case of a vessel that has never been used for survey work before, a safety audit should be considered mandatory. If large amounts of physically heavy survey equipment are to be installed on a small vessel the ship's stability calculations may need to be reworked by a naval architect. To give one example, the vessel waterplane coefficient of form gives the sinkage in tons per inch when weights are added to the hull. This information should be known onboard at the start of mobilisation.

A major limitation on any survey vessel is noise caused by the ship's propellers. It has even been the case that a purpose-built survey vessel had a

> **Box 1.4**
> **What happens when a survey instrument room does not have a dedicated power supply**
> Once upon a time there was a site survey operation that was thoroughly competent. It lacked one thing only to make it absolutely superb. The instrument room power supplies were provided from the survey ship's generators, not from a dedicated instrument room generator. Everything worked wonderfully until Sunday morning when the electronic navigation system stopped working. The engineers worked on it for hours and never found a fault. In mid-afternoon it suddenly started working again. The next Sunday the same thing happened. Then on the third Sunday, the author and the party manager had a long hard look at the instrument power supply and monitored not just the power supply voltage going to the instrument room but the power supply frequency which was well down. It turned out that the galley was all-electric and supplied from the same generator as the instrument room power supply. On Sunday all six heating plates plus four ovens were used to prepare the Sunday lunch. This was not the case on weekdays when the comestible arrangements were less permissive. The navigation system was particularly sensitive to voltage power supply frequency fluctuations, hence the 'Sunday lunch effect'. Needless to say the curtailment of Sunday lunch preparations did not make the author popular.

main engine configuration that produced very high levels of streamer noise that terminated its career as a survey boat. Most survey vessels use constant-speed variable-pitch propellers, though the last few years have seen a move to fixed-pitch variable-speed propellers.

Modern developments in ship propulsion include axial thrusters. These units are usually podded which means the electric drive motor is positioned in a pod below the ship's hull, with the electric drive directly to the propeller. With conventional thrusters a significant amount of noise is generated by the right angle bevel gearing of the drive. When a vessel is taken up for site survey work, the type of propulsion should be known and the contractors asked to provide evidence of the vessel's acoustic noise signature. Diesel–electric propulsion is preferable to pure diesel propulsion since the prime mover (the diesel) is removed from the drive system (electric motors).

Figure 1.7 shows an outline vessel schematic. The approximate position of the streamer reel is indicated but close attention has been paid to the position of the navigation reference point, the echosounder transducer position and the gyro position. It should be noted that most survey vessels have two gyros, a ship's main gyro and a survey system gyro. The exact measured position of these survey elements should be established in port before the start of every survey.

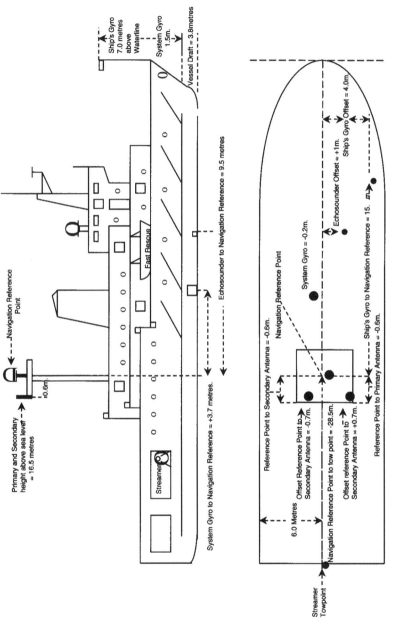

Figure 1.7 Typical vessel schematic. Particular attention has been paid to navigation, gyro and echosounder offsets.

1.18 Reporting

Contractors who perform high resolution site surveys usually produce a report for the oil company or other operating authority within 30 days of the survey completion. Shallow gas accumulations should always be clearly identified. All information used in the interpretation of the data and the assessment of shallow gas potential should be included. Interpretation of shallow gas seismic data may also be useful for guide base foundation studies, conductor/casing setting studies and the detection of difficult drilling zones. The depth and lateral extent of all features should be given. A prediction of lithology should also be included. For foundation and anchoring purposes the depth and lateral extent of all relevant soils should be included. Bathymetry maps should be constructed and reduced to least astronomical tide (LAT). Seabed sediments and obstructions should be described in the report and mapped in their correct positions.

The oil company will have certain statutory reporting requirements in relation to the country in which the survey has been conducted. In the case of the UK sector of the North Sea, a reproducible transparency of each final processed seismic section together with a transparency of the shotpoint interval map is usually required. Sidescan sonar records with the associated track charts are also required. A copy of the contractor's site survey report should be available.

The consulting engineer, variously referred to as the client representative, quality control supervisor, or in American parlance the 'bird dog', appointed to a particular survey will provide a daily report whose format will be agreed with the oil company or operating authority in advance of the survey. Typically this will include all the contractor charged kilometres for that particular day, if the contract is on a per kilometre rate, or all charged time if the contract is arranged on a operational time basis. In fact, most site surveys work on a charged operational time basis, though this is not axiomatic. The daily report will also provide an analysis of time on prospect for that particular day, clearly differentiating between charged and non-charged time, and a brief description of the day's events.

At the end of survey the 'bird dog' QC draft report may be ready when the survey vessel reaches port. If this is not the case a draft report should be ready within 2–3 days. This will allow the oil company to check invoices and to make an initial assessment of the data. The final report should be ready for presentation within four weeks.

The format of a site survey report can be varied to meet almost any particular client requirement. A typical site survey report would adopt the following format.

1. *Introduction.* This will state in general non-technical terms the nature of the survey, the survey area and programme, the parameters of the survey, the contractors involved and the outcome of the survey. Typically this will be a single page introduction.

2. *Conclusions and recommendations.* This will state in brief terms the conclusions arrived at by the QC supervisor. These will be stated in general operational terms and in terms of specific instrumentation. An assessment and perhaps an initial interpretation of the data can be made. Any recommendations will be included. Typically the conclusions and recommendations will be two or three pages in length.
3. *Survey and recording parameters.* This will state in technical terms the geotechnical parameters of all the survey systems and the geodetic parameters for the navigation systems.
4. *Prospect.* This section of the report will give a detailed description of the survey operation and deal with matters such as weather experienced on site, fishing activity on site, seismic interference on site and any other matters of operational interest.
5. *Survey analysis.* This will give the exact production (charged hours), the production seismic kilometres and all downtime hours charged to the oil company. A complete analysis of time spent on prospect will be given. This section of the report will allow costs to be checked against contractors' invoices.
6. *Digital seismic survey systems.* This section will give a technical description of the instruments used, all quality control tests and calibrations run on them and the operational performance of the systems, including all faults and fault remedies.
7. *Analogue seismic survey systems.* This section will give a technical description of the different analogue instrument packages used, all tests and calibrations performed and all faults and fault remedies.
8. *Positioning systems.* This section of the report will deal with the positioning systems, the integrated navigation system if carried, or the method of integration between the positioning system and the seismic systems. Again, all faults and fault remedies will be described.
9. *Appendices.* Line logs, adopting an Excel spreadsheet format, will cover the digital seismic acquisition, the analogue seismic acquisition and all navigation acquisition. Drawings of all offsets, laybacks, system overviews in the form of schematics, etc., will be provided. Programme maps will also be provided as necessary. A detailed daily diary will describe all events on a day-by-day basis with all times assigned to various operations during each day.

Obviously the report format will vary according to the survey. Very short surveys of limited extent may have several of the sections described above integrated. Exceptionally long surveys may have extra sections to cover additional systems or extensive operations descriptions. Particular oil companies may have their own report writing format. Site survey reporting is almost infinitely flexible and based, on Word and Excel computer facilities, the right format for a particular survey can be arrived at.

Suggested background reading

Background reading for site surveys covers a huge range of published literature. The following is a selection that may help the reader in further researches into site surveys.

Admiralty List of Radio Signals, Volume 8, 1998/99, UK Hydrographic Office, London.
Admiralty Manuals of Hydrographic Survey, Volumes 1 and 2, 1998/99, UK Hydrographic Office, Taunton.
Ardus, D.A. (1980) *Offshore Site Investigation*, Graham and Trotman.
Belderson, R.H., Kenyon, N.H., Stride, A.H. and Stubbs, A.R. (1972) *Sonographs of the Sea Floor*, Elsevier Publishing Corporation.
Denness, B. (1980) *Seabed Mechanics*, Graham and Trotman.
Dobrin, M.B. (1976) *Introduction to Geophysical Prospecting*, McGraw-Hill Book Company.
Laurila, S.H. (1976) *Electronic Surveying and Navigation*, John Wiley and Sons.
Leick, A. (1990) *GPS Satellite Surveying*, John Wiley and Sons.
Loweth, R.P. *Manual of Offshore Surveying for Geoscientists and Engineers*, Chapman and Hall, London.
Mcquillin, R. and Ardus, D.A. (1977) *Exploring the Geology of Shelf Seas*, Graham and Trotman.
Mcquillin, R., Bacon, M. and Barclay, W. (1979) *An Introduction to Seismic Interpretation*, Graham and Trotman.
UKOOA Guidelines for the Conduct of Mobile Drilling Rig Site Surveys, Volumes 1 and 2, 1987, UKOOA, London.

References

1. Buxton, I. (1978) *Big Gun Monitors*, World Ship Society and Trident Books, Tynemouth, Tyne and Weir, p. 197.
2. McQuillin, R., Bacon, M. and Barclay, W. (1979) *An Introduction to Seismic Interpretation*, Graham and Trotman, London, p. 2.
3. Hackmann, W. (1984) *Seek and Strike, Sonar, Anti-submarine Warfare and the Royal Navy 1914–1954*, Her Majesty's Stationery Office, London, pp. 228–231.
4. Telford, W.M., Geldart, L.P., Sheriff, R.E. and Keys, D.A. (1976) *Applied Geophysics*, Cambridge University Press, p. 334.
5. Arthur, J.C.R. (1979) 'The application of high resolution multi-channel seismic techniques to offshore site investigation studies', *Offshore Site Investigation, Proceedings of a Conference Held in London*, March 1979, p. 77.
6. Mcquillin, Bacon and Barclay, op.cit., p. 9.
7. *UKOOA Guidelines for the Conduct of Mobile Drilling Rig Site Surveys*, Volume 2, p. 33.
8. Ibid., p. 33.
9. Ibid., p. 37.

2 High resolution digital site survey systems

2.1 Digital recording systems

In the days of analogue recording the maximum dynamic range of seismic recording instruments was in the range 20–25 dB. The early digital recording systems used a fixed gain with about 80 dB of total gain retention. Floating-point digital recording systems give up to 100 dB amplitude retention.[1] Keeping large variations in signal gain is essential for relative amplitude seismic records and this is the basis of bright spot analysis and high resolution site surveys.

The first attempt at using a digital recording system for site survey work was a Texas Instruments 960A computer, allied to a number of dedicated interfaces. In its original form this was the system used by Fairfield-Aquatronics in the early 1970s. The author of this book received his first initiation into the world of site surveys as a systems engineer with Fairfield-Aquatronics working on this system. This precursor system was superseded by conventional 2D recording systems such as DFS III (Digital Field System Mk III), then DFS IV and eventually DFS V. Other systems used for site survey work included the CGG HR 6300 (Compagnie Générale Geophysique) system, Quantum DAS-1A (Digital Acquisition System Mk 1A) and MDS-10. These systems should all be considered obsolete, though some DFS V systems remain in existence. Systems in operation today include the multiplexer-based Sercel SN 358 DMX system and sigma–delta systems such as TTS-2, OYO Geospace DAS-1A and the Geometrics Strativisor NX seismic recording systems. The last two named systems are suitcase portable. Figures 2.1, 2.2 and 2.3 show outline schematics of the DFS V, TTS-2 and Geometrics Strativisor systems.

Older systems such as DFS III and Quantum DAS 1A (not to be confused with the OYO DAS-1A) were binary gain ranging systems. These systems used gain steps of 6 dB. Systems such as DFS IV and DFS V were quaternary floating-point systems and used gain steps of 12 dB. DFS V became the industry standard, but for site surveys there is no particular advantage in a system that uses such large gain steps. The dynamic range of the acoustic signals used for site surveys is relatively smaller than for 2D surveys and there were always some advantages in using a recording system with 6 dB gain steps. In DFS IV and DFS V, floating point is actually a notation system in

High resolution digital site survey systems 25

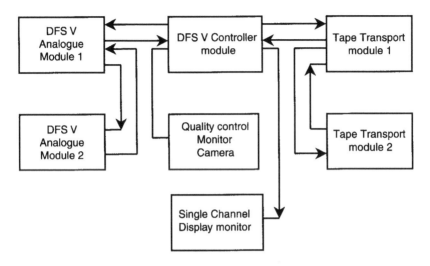

Figure 2.1 DFS V outline schematic.

which a numerical value is expressed as a mantissa and an exponent floating point. Notation is a convenient method of expressing numbers which vary over a very large range.

Any floating-point system examines the amplifier output at each sample and sets the gain to the highest value possible without overdriving the converter or overloading any amplifier gain stage. The data sample is digitised after the gain is set. The floating-point amplifier will, in general, have several different gain values during a cycle of the signal. This renders the data word meaningless without the gain word. The converter output is treated as a mantissa (data value). The floating-point amplifier supplies the exponent. An amplifier gain of unity (exponent of 0) means the converter output is four times the true signal level. Moving the point two places to the left (-1 for a quaternary exponent) gives the true signal level. A 3-bit quaternary gain word provides a gain range of 84.28 dB and this is the theoretical dynamic range of this type of converter.[2]

The early digital recording systems recorded multiplexed data. This was data sampled on a time-sequential basis. The limiting factor on fast sample rates was for many years dependent on multiplexer switching speeds. Demultiplexed data is trace sequential data. The recording formats used were the Society of Exploration Geophysicists (SEG) formats, SEG A, SEG B, SEG C and SEG D. SEG A is now virtually extinct. SEG B is an instantaneous floating-point multiplexed format and SEG C is a full floating-point multiplexed format that uses approximately 60 per cent more tape than SEG B. SEG D is much more modern and can cater for both multiplexed and demultiplexed formats. SEG Y is a demultiplexed format.[3]

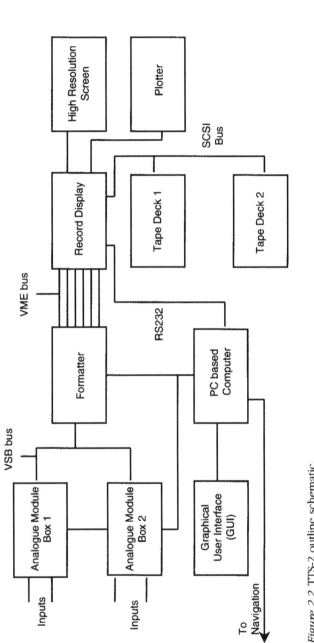

Figure 2.2 TTS-2 outline schematic.

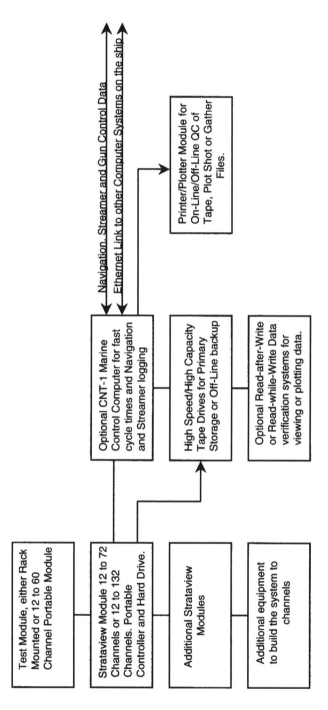

Figure 2.3 Geometrics Strativisor marine seismic recording system.

Since DFS V represented the industry standard for so long, a relatively detailed explanation is given here and hopefully provides some historical context to these systems. It has already been stated that DFS V is a multiplexer-based system, as is the Sercel SN 358 system. DFS V suffers from a severe limitation in that 1 s of cycle time is needed between shots. This precludes its use for 6.25 m shotpoint intervals with the associated 2.5–3.0 s firing rates. The Sercel system is much more modern and does not suffer from this defect even though it is a multiplexed system. If a streamer group interval of 12.5 m is used and this is entirely typical, then a 6.25 m shotpoint interval will generate full-fold coverage. This will always be the case, since the sub-surface coverage is half the surface coverage. A 12.5 m shotpoint interval combined with a 12.5 m streamer group interval will give only 50 per cent of full-fold coverage, almost invariably the case with DFS V (see Figure 1.3 already referred to).

TTS-2, OYO DAS-1A and the Strativisor NX system are all sigma–delta systems which should be considered the logical successors to multiplexer-based systems, mainly because of their ability to handle 6.25 s shot intervals allied to 2.5–3.0 s digital record lengths. Oil companies should understand that the choice of an older digital recording system may result in 50 per cent coverage, whereas the use of newer sigma–delta systems almost always allows the recording of full-fold data.

Typical minimum digital seismic acquisition parameters for high resolution surveys might be as follows:

- Forty-eight-channel digital acquisition.
- Streamer length of 600 m.
- Streamer group length of 12.5 m (48 traces).
- Shot interval of 6.25 m, for full-fold coverage.
- Source peak-to-peak output > 5 bar m.
- Source and streamer should be maintained at a uniform depth, typically 3 m ± 5 m.
- The separation between the source and the centre of the streamer near trace (dynamic offset) should not exceed half the minimum water depth of the survey area.

2.2 Texas Instruments digital field system (DFS) Mk V

A DFS V usually consists of a controller module, two analogue modules and two tape transports. Figure 2.4 shows the DFS V at the centre of a complete seismic recording package that includes an airgun array and hydrophone streamer. The analogue module performs three functions. First, it accepts the analogue data and converts it to a digital format. Second, it performs auxiliary functions not directly related to data acquisition, such as time break lengthening. Third, it performs test and calibration functions. Digitised information is transmitted to the controller module where the data is formatted and transmitted to the tape transport module. The controller module provides all the timing commands to

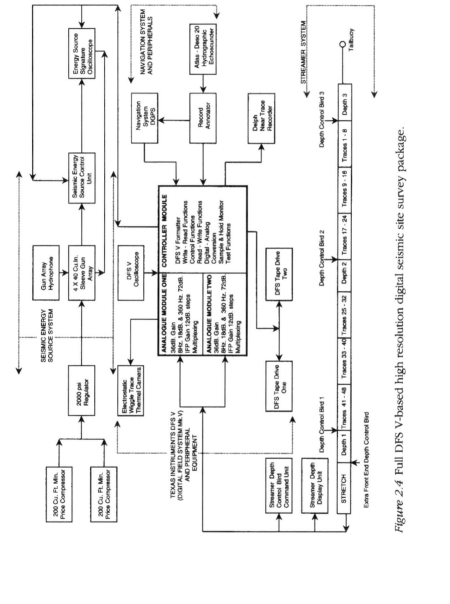

Figure 2.4 Full DFS V-based high resolution digital seismic site survey package.

the analogue and transport modules. These consist of channel addresses for sampling. The control module also provides IFP (independent floating point) gain ranging, up and down signals and status signals. Before addressing an analog input channel, two time intervals must occur. The first interval (first start of scan) is an address providing transmission of filter and gain settings which also resets the IFP amplifier. The second time interval (second start of scan) short circuits the IFP input and therefore provides any zero offset voltage levels. Upon completion of both time intervals the address sample sequence is continuously generated until the 'end of scan' signal is received.[4]

In sum, the controller module controls the operation of the analogue module and performs four primary functions. It addresses the channels to be sampled, it commands the sending of status and zero offset data, it controls whether the gain ranging amplifier automatically selects its gain or operates at a gain set by the operator. Finally, it controls the source of input to the analogue-to-digital converter.

A major advantage of DFS V over older systems such as DFS IV is the mnemonic display fault isolator, which displays as many as sixty-six fault conditions visually on the controller module. These fault conditions have two classes designated primary and secondary. Primary faults can be overridden but secondary faults cannot. The mnemonics are listed on a card according to their address sequence. It has already been stated that a major disadvantage of DFS V is the cycle time between shots, which makes data acquisition at a shotpoint interval of 6.25 s difficult or perhaps impossible. This occurs in the following manner. The navigation system issues a start command to the controller module, which issues a system start command to the tape transports. The system has a 900 ms delay period, at the end of which the blast command is given. This delay allows the transport to reach the correct speed and provides a period in which the file header is written. The problem of the recycle time lies with the 900 ms delay, which cannot be circumvented due to the finite time required for internal signal processing. Figure 2.4 shows a schematic of a typical site survey digital recording system and streamer.

Typical DFS V recording parameters might consist of the following:

- Sample rate: 1 ms (0.5 ms for some applications)
- Record length: 2 s, data to 1500 m below the seabed (typical)
- Low cut filter: 27 Hz
- High cut filter: 256 Hz
- Gain mode: IFP
- Format: SEG 'B' is standard

2.3 Sercel SN 358 digital recording system

The Sercel SN 358 is a conventional seismic data acquisition system with the ability to handle very fast sampling rates and large numbers of input channels. Configurations available extend from eight channels at a 1/8 ms sampling rate,

High resolution digital site survey systems

to 132 channels at a 2 ms sampling rate. Sercel now offer an extended SN 358 system which can record up to 246 seismic channels at a 2 ms sampling rate. This system records data in a demultiplexed format, SEG D, $2\frac{1}{2}$ bytes per word.[5]

An SN 358 system will usually consist of a master analogue unit which consists of twenty-four seismic channels with four auxiliary channels, test circuits, the IFP amplifier and the analogue-to-digital converter. A number of slave analogue units will also be used. A total of three slave units can be connected to a master analogue unit. Each slave unit houses up to thirty-six seismic channels. There will also be a logic unit which contains the central microprocessor. This unit controls all the peripherals to format the tape decks, control the tape decks, read back the tape, perform AGC functions and to perform all the computations needed for the tests. The digital-to-analog converter and sixty-four analogue playback filters and amplifiers are also controlled by the microprocessor. There will also be a tape transport, usually a single density 1600 bpi PE or perhaps a 1600 bpi PE/6250 bpi GCR unit. A control panel is also necessary to gather the input and display features to control the entire system. Peripheral equipment can include a line printer, a conventional stacker and a correlator stacker. Optional equipment includes a dual tape transport unit and a demultiplexer memory unit. This unit records the demultiplexed data in SEG D demultiplexed format already referred to.

The input to the system can be transformer coupled with fixed channel gains at 2^5 or 2^8. A transformerless input is available for marine operations with fixed channel gains of 2^4 or 2^7. A low cut filter at 8, 12.5 or 20 Hz can be used with slopes at 18 or 36 dB/octave, while the anti-alias filter is a Cauer type altered according to the sample rate at 77.2, 154.5, 308.8, 618 and 2470 Hz. Twelve multiplexer circuits are grouped on the SN 358 multiplexer board. One single IFP amplifier is used unless the DMX system is in use, in which case two IFP amplifiers are necessary. A single 15-bit analogue-to-digital converter is all that is necessary, unless again the DMX system is being used in which case two analogue-to-digital converters are used. Recording parameters for a typical survey will resemble those for a DFS V system.

Typical recording parameters might be as follows:

- Sample rate: 1 ms (0.5 ms for some applications)
- Record length: 3 s, data to 1500 m below the seabed (typical)
- Low cut filter: 8, 12.5 or 20 Hz, 18 or 36 dB per octave roll-off
- High cut filter: 77.2, 154.4, 308.8, 618, 1235, 2470 Hz (308.6 Hz at 1 ms)
- Gain mode: IFP
- Format: SEG 'D' is standard

2.4 TTS-2

The whimsically named TTS-2 system, TTS being the initials of the original designers, is a much more modern digital recording system used for site surveys. TTS-2 is a sigma–delta system which can record 128 channels at a

1 ms sample rate, with a bandwidth of over 400 Hz. Sample rates can be varied from 2 ms down to 0.25 ms. A 24-bit analogue-to-digital converter allows recovery of signals from very high levels of noise. Data conversion is by a 24-bit delta–sigma modulation technique with digital filtering. These converters eliminate the need for IFP (independent floating point) amplifiers, sample-and-hold circuits and analog anti-alias filtering. They are inherently linear and the result is a digital recording system with very low distortion. This gives a near ideal pulse response.[6] Figure 2.5 shows the full TTS-2 schematic.[6]

Twenty-four-bit converters are a huge advance over the older 16-bit analogue-to-digital converters. There is a large increase in instantaneous dynamic range, that is, the ability to record high and low level signals simultaneously. A good example of how this works in practice is a high level of low frequency noise which tends to push down the gain of the IFP amplifiers and thus the desired signal amplitudes. Post-filtering can easily and effectively remove the high levels of low frequency noise.

Another major advance associated with TTS-2 is the use of a single analogue-to-digital converter per channel, usually referred to as delta–sigma sampling. Analogue multiplexers are no longer required and this removes the main sources of interchannel crossfeed, sample skew and most common sources of noise and distortion.

This system does away with the cycle time between records and makes a 6.25 m shot interval, usually a requirement on contemporary site surveys, a much more practical proposition (already mentioned above). Recording parameters might be as follows:

- Sample rate: 1 ms (0.5 ms for some applications)
- Record length: 2 s, data to 1500 m below the seabed (typical)
- Gain mode: IFP
- Format: SEG 'B' is standard
- Base gain: 30 dB
- Low cut filter: 8 Hz, 18 dB/octave
- High cut filter: 411 Hz, 72 dB/octave
- System set-up: 96 channels, four auxiliary channels
- Tape deck format: SEG-D 8015, 20-bit demultiplexed

In practical engineering terms the streamer is directly coupled to the instrumentation system. There is a front-end multiplexer on each channel which selects a seismic input, test oscillator input or terminated input. A value of base gain is then selected by the operator. There are two gain settings available at 24 dB or 30 dB. The 30 dB value is probably more typical than the 24 dB value. The input signals pass to a control gain amplifier. This pre-amplifier gain is software controlled and is set according to the level of signal input. Essentially the deeper the water column, the greater the level of signal attenuation and the greater the value of gain required. The pre-amplifier gain operates in 6 dB steps, so with 30 dB of base gain the pre-amplifier gain steps are 36 dB, 42 dB,

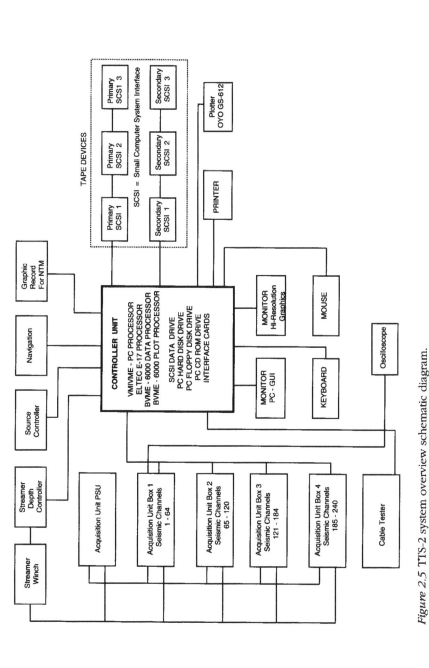

Figure 2.5 TTS-2 system overview schematic diagram.

34 *High resolution site surveys*

48 dB and 54 dB. After gain has been applied the input signals go to a buffer amplifier and then to a low cut filter. The low cut amplifier frequency is typically 8 Hz with a slope of 18 dB per octave. The input signals are then digitised by the 24-bit sigma–delta oversampling converter and a high cut filter is applied. For a sample rate of 1 ms the filter is 411 Hz.

The advantages of this system, already mentioned, over other older systems are a much greater instantaneous dynamic range within the seismic frequency band and the ability to use digital anti-alias filters which are far superior to their analogue counterparts. The digitised data then goes to a 32-bit shift register and a VME bus. The data is formatted in SEG-10 demultiplexed format, with the data processor unit actually performing the demultiplex. Finally the data is written to the tape cassettes using a SEG-D 8015 format on 3490 IBM compatible cartridge drives. For quality control purposes an OYO plotter might be used for individual shot monitors and as a near trace monitor. The TTS-2 has a maximum capability of 128 data channels and four auxiliary channels. Four channels are contained on each data board.

2.5 OYO DAS-1A

The OYO Geospace DAS-1A digital recording system is designed specifically for site survey work. It is the first sigma–delta recording system designed to be air transportable for site survey operations. Figure 2.6 shows a full schematic of this system. The entire system is built into 'suitcase' boxes for air transportation as excess baggage. The system can operate as few as twenty-four channels and as many as 144. Very fast sample rates are possible, from 31.25 μs to 4 ms. In common with other newer digital recording systems the DAS-1A uses sigma–delta analogue-to-digital conversion with 24-bit resolution. This gives a very high system dynamic range (132 dB) and a large potential variation in management software, data recording, data processing and quality control monitoring.[7] The recording system is allied to a GCCC digital streamer, typically a 96-trace mini-streamer with a 12.5 m group interval. With a shotpoint interval of 6.25 m, 96-fold coverage is possible with a 2.0–3.0 s digital record.

Data from the seismic channels is amplified, filtered and converted to 18-bit digital data by the sigma–delta A/D converter and sent to the data acquisition controller board (DAC) for processing by the DSP central processing unit (CPU) with a filter algorithm for digital filtering and oversampling. The 24-bit output from each digital filter operation represents one data channel. This data is sent to the data handling board for sorting, processing, demultiplexing and temporary storage as 32-bit data in the data memory (DRAM).

The 486 CPU then processes the data and transfers it to the hard disk and thermal plotter. The data is also recorded on a 3840 format tape cartridge.

The sigma–delta converter can be referred to as an oversampling converter. A sigma–delta converter quantises an analogue signal with very low resolution (1 bit) and a very high sampling rate (about 1 MHz). With the use of

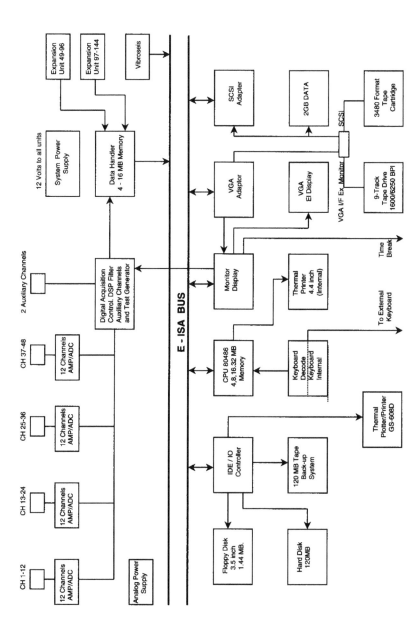

Figure 2.6 OYO DAS-1A schematic diagram.

36 *High resolution site surveys*

oversampling techniques and digital filtering, the sampling rate is reduced to approximately 16 kHz and the resolution is increased (18 bits). The sigma–delta converter can be considered as one where an analogue signal is sampled at one rate to produce digital data. The data is then resampled at another rate. The main disadvantage of older systems such as DFS V is avoided, namely the limitation placed on the system by multiplexer switching times. Typical recording parameters are given below.

- Sample rate: 1 ms
- Coverage: 96-fold
- Record length: 2.5 s at a firing rate of 3.0 s
- Low cut filter: 0 Hz, 30 dB/octave
- High cut filter: 375 Hz, 70 dB/octave
- Tape drives: Two 3480 cartridge drives
- Format: SEG 'D'
- On-line QC system: Lookout RWW 14 QC
- Near trace recorder: Delph

2.6 Geometrics Strativisor NX seismic recorder

In common with the OYO DAS-1A the Strativisor NX system is suitcase portable. In its suitcase portable form it can provide 12–72 seismic channels, which can be extended to 132 channels if a rack mounted system is adopted.[8] Even the portable suitcases are designed to be rack mounted. Addition of further modules can boost the system to 600 channels. For small surveys each module can be operated as a stand-alone seismic recorder without an external controller (see Figure 2.3 already referred to).

The R-Series engineering system has an exceptionally wide bandwidth at 14 kHz, while the RX high dynamic range system has only 0.0003 per cent distortion. The RX system uses the Crystal Conductors sigma–delta analogue-to-digital converter. In common with the TTS2 and DAS-1A systems there is effectively no cycle time between shots, so that shotpoint intervals of 6.25 m are possible, for full-fold data coverage. Data displays are capable of the following:

- Shot display;
- Near trace (common offset) gather;
- Log file showing tape status, acquisition parameter changes and channels exceeding noise specifications;
- Noise bar graphs with user definable thresholds;
- Gun energy bar graphs for gun performance monitoring.

The controller system runs under the Windows NT operating system. Quality control monitoring is fairly extensive and there are real-time bar graphs for

High resolution digital site survey systems 37

monitoring streamer noise. This can be done between shots to avoid shot noise. Each shot can be displayed in a window with AGC or manual scaling even at the fastest cycle times. A near trace gather is available for continuous data monitoring on-line. An on-line brute stack can be produced for quality control and initial data interpretation. This brute stack does not replace a full processing sequence. It applies a user-defined velocity function to the shot gather, sorts the traces into a midpoint gather and adds stacking data between shots. A semblance analysis tool is used to determine the stacking velocity function. This gives the crude brute stack.

The general characteristics of the system are as follows:

- Channels: 12–72, extendible to 132 channels
- A/D conversion: 24-bit A/D using crystal semiconductors sigma–delta
- Dynamic range: 144 dB (theoretical), 125 dB (instantaneous measured) at 2 ms, 0 dB
- Distortion: 0.0003 per cent at 2 ms, 0.3–206 Hz
- Bandwidth: 0.3–1.7 kHz
- Common mode rejection: >-110 dB at ≤ 100 Hz, 48 dB
- Crosstalk: -120 dB at 100 Hz, 0 dB
- Noise floor: 0.18 V, RFI at 2 ms, 0.3–206 Hz
- Stacking trigger accuracy: 10 V peak-to-peak
- Input impedance: 20 kΩ 0.02 µF
- Pre-amplifier gains: 0, 24, 36 and 48 dB
- Anti-alias filters: 82 per cent of Nyquist, down 130 dB
- High cut filters: Out, 250, 500 or 1000 Hz
- Low cut filters: 10, 15, 25, 35, 50, 100, 140, 200, 280, 400 Hz, 24 or 48 dB roll-off
- Sample rate: 0.25, 0.5, 1.0, 2.0, 4.0, 8.0, 16.0 ms
- Data formats: SEG-D 8048 and 8058

2.7 Digital systems tests

Before the start of survey a full set of digital instrument tests should be performed and subjected to computer analysis either onboard the survey vessel or in a recognised processing centre. This will usually be a mandatory requirement by the organisation contracting the survey. Some oil companies have their own suite of tests for a particular system. Most survey companies have their own set of tests which they will perform either back-to-back with the oil company or independently. Both parties to a survey may agree to use the manufacturer's set of standard instrument tests.

Two sets of instrument tests are included here, one related to a multiplexer-based recording system, DFS V[9] and another based on a sigma–delta recording system, TTS-2.[10] The tests discussed relate to families of systems. DFS V and the Sercel SN 358 belong to the multiplexer family, while TTS-2, OYO

> **Box 2.1**
> **Things that can go wrong with instrument tests**
> In the days before onboard processing only paper records were available to assess the status of the digital recording system. In the case of the DFS filters, the paper records indicated that all channels were behaving in the same way at the same time but the paper records did not indicate the frequency of the filters. A contractor set up a system with the wrong low cut filter settings. These were determined only by which way round the filter chips were positioned on the printed circuit boards. This error came to light half way through the survey. Needless to say everyone blamed everyone else, but this sort of mistake shows how important onboard computer processing has become for avoiding MCUs (monumental cock-ups).

DAS-1A and the Geometrics Strativisor belong in the sigma–delta family of systems. Tests can be divided into static and dynamic tests. Static tests include zeroing the analogue module, test signal oscillator and AC meter calibration, gain setting, ohmmeter and leakage meter calibration, tape deck speed check, tape deck skew checks, float amplifier zero, multiplexer zero, gain constant check, header check and a system polarity check.

DFS V instrument tests

System noise test

To evaluate the input noise level of the system the streamer input lines are replaced by $2\,k\Omega$ resistors. A known calibrated sine wave is then compared to the noise record of the terminated channel. Generally speaking the system noise should be less than $0.5\,\mu V$.

Dynamic range tests

The converter dynamic range is conducted at low amplifier gain so any noise is assumed to be converter noise rather than amplifier noise. The test consists of supplying the input with successively lower signals and providing a correspondingly higher gain in playback. Each test signal step reduces the signal level by $4:1$, which is 12 dB.

Crossfeed isolation tests

The crossfeed isolation test measures the amount of feed from one channel to any other channel. Basically it consists of feeding the maximum signal short of saturation into one of the channels and observing the output on the other channels when played back at high gain. All the unexcited channels should have their inputs by a 2 kΩ resistor. Normally all the odd number channels are driven and the even numbers are terminated. The test is then repeated with the odd number channels terminated and the even number channels driven.

Converter calibrate checks

As a check on the operation of the A/D converter, a set of voltages both plus and minus, corresponding to the particular bit weights of the A/D, can be fed in and then observed to see how closely the bit weights correspond to their theoretical values.

Exponential oscillator, IFP amplifier AGC tests

The independent floating-point (IFP) oscillator test is designed to exercise the amplifier system across its entire input signal range. The test is made using an exponentially decaying sine wave test signal. The automatic gain control (AGC) system in the playback mode is also exercised by this test. The playback during the initial recording is made in the float mode to emphasise the gain steps which are made by the floating-point amplifier. This produces a record with little resemblance to the input waveform. The test tape record is then played back in the defloat mode which should show an exponentially decaying waveform. Finally, the record should be reproduced in the AGC mode. This test shows, therefore, a decreasing amplitude as the AGC system brings the signal down from its initial amplitude of over 70 per cent of full scale down to the set level of the AGC system.

Gain ranging checks

The test signal attenuator and the gain steps of the floating-point amplifier are dependent on very good resistors, as are the pre-amplifier and the filters on the filter multiplexer card. The gain linearity check compares these three elements, namely the test signal attenuator, the floating-point amplifier and the filter multiplexer together. If any one link in the chain develops a fault, it should show as a non-linearity. All the signals from the test should be nominally the same. However, when the gain constant is maximum the effect of the rated noise on the system is sufficient to raise the reading by about 10 mV. What is important in this test is not the absolute accuracy but the constancy of the readings which should all stay within 1 per cent from channel to channel. A series of readings taken on one particular channel with varying inputs and gains should hold constant within 0.5 per cent.

Pulse tests

By exciting the system with a pulse, the filter response may be evaluated. Under field conditions only obvious faults can be detected. It is essential that

40   *High resolution site surveys*

the data from this test be checked on a computer system such as Promax. This allows a Fourier transform to be performed on the impulse response to determine the frequency response of the filter. The computer can also make accurate determinations of the zero crossings and the peaks by comparing the various channels to each other with respect to a time delay.

Skew tests

Skew tests measure the tape deck recording head alignment to the recording tape itself. Both read and write skew should be measured.

Test analysis

The results of the system tests should be analysed onboard using a Micromax or Promax system. A proportion of these tests are repeated on a daily basis, to continuously check the digital instrumentation during survey. Similar tests for the Sercel SN 358 system would include much of the above with the inclusion of line tests for continuity and leakage with the results for each channel displayed on the control panel. Logic tests are automatically performed each time the system is switched on and these check the main parts of the system. A magnetic tape dump test and a tape transport utility test are also performed automatically.

Daily tests are often an abbreviated form of the pre-survey tests, agreed between the oil company and the contractor.

Similar tests for sigma–delta recording systems such as TTS-2 and OYO DAS-1A should include the following.

TTS-2 system tests

System dynamic range

For a fixed gain system the dynamic range is the same as the signal-to-noise ratio and will vary with overall fixed gain. The contractor's typical quoted figures usually show the ratio between the maximum input voltage and the average noise, referred to the input value. The normal measured values are the maximum oscillator voltage and the measured input noise level, which gives values approximately 1 dB less than those shown in Table 2.1.

Instrument noise and DC offset (Tables 2.2 and 2.3)

The low cut filters should be out for this test and a source resistance of $2000\,\Omega$ used. Recording or calibration tests should not be made within 30 min of initial switch-on to allow the analogue-to-digital converters (ADCs) to stabilise. The ADC offset is the dominant factor in the specification. They have an inherently high internal offset (which has no effect on dynamic range). This is corrected by an internal mechanism within the ADC which calculates and compensates for the offset. A further correction can be performed on the digital data using

Table 2.1 TTS-2 system dynamic range

Anti-alias filter (Hz)	Sample rate (ms)	Dynamic range system average (dB) for total fixed gain (dB)				
		24	30	36	42	48
206	2	120	117.5	114	108	102
412	1	116.5	114	111.5	105.5	99.5
824	$\frac{1}{2}$	112	110.5	108	103.5	97.5
1652	$\frac{1}{4}$	101	101	100	97.5	93.5

Table 2.2 Instrument noise, system average

Anti-alias filter (Hz)	Sample rate (ms)	Noise, system average (µV referred to input) for total fixed gain (dB)				
		24	30	36	42	48
206	2	0.6	0.4	0.3	0.3	0.3
412	1	0.9	0.6	0.4	0.4	0.4
824	$\frac{1}{2}$	1.5	0.9	0.6	0.5	0.5
1652	$\frac{1}{4}$	5.3	2.7	1.5	1.0	0.8

Table 2.3 System noise, absolute maximum

Anti-alias filter (Hz)	Sample rate (ms)	DC offset, absolute maximum (µV referred to input) for total fixed gain (dB)				
		24	30	36	42	48
206	2	3.0	1.5	1.0	1.0	1.0
412	1	4.0	2.0	1.5	1.5	1.5
824	$\frac{1}{2}$	4.0	2.0	1.5	1.5	1.5
1652	$\frac{1}{4}$	16.0	8.0	4.0	4.0	4.0

a digital filter. This can be done either manually or automatically on-line. This filter is not applied to test data to avoid masking errors. Table 2.4 gives the offsets after correction, but without the digital offset filter.

In the OYO DAS-1A system the DC offset and noise test both analyses the signal input and calculates the DC offset/system noise. For this test the signal generator is turned off and the channel inputs are grounded. A set of 2048 samples from each channel is collected and the DC offset and noise are calculated for each channel.

Interchannel crossfeed

Interchannel crossfeed should be measured with full scale signal on alternate channels, undriven channels terminated. A test frequency of 7.8–125 Hz

42 High resolution site surveys

Table 2.4 DC offsets

Anti-alias filter (Hz)	Sample rate (ms)	DC offset, absolute maximum (µV referred to input) for total fixed gain (dB)				
		24	30	36	42	48
206	2	3.0	1.5	1.0	1.0	1.0
412	1	4.0	2.0	1.5	1.5	1.5
824	$\frac{1}{2}$	4.0	2.0	1.5	1.5	1.5
1652	$\frac{1}{4}$	16.0	8.0	4.0	4.0	4.0

should be used. The result should be better than 90 dB of crossfeed isolation at all filter settings.

In the OYO DAS-1A system the alternate channels (odd/even) with a pre-amplifier gain of 24 dB are connected to the signal generator and the others (even/odd) at 48 dB are connected to ground. The test collects 2048 samples on the grounded (even) channels and the connections are then reversed and data is collected for the odd channels. The crosstalk values in decibels are then calculated.

Distortion

The TTS-2 system exhibits very low distortion at normal levels of signal input but care must be taken to ensure that any external test equipment used to check distortion does not affect the result. The harmonic distortion can be measured against amplitude at base gains of 24 and 36 dB. Finally a common mode rejection ratio versus frequency value can be calculated. The maximum permitted distortion allowable is usually 0.005 per cent.

In the OYO DAS-1A system the total harmonic distortion is measured by taking all channels with 24 dB of pre-amplifier gain and connecting them to a signal generator which generates 98 per cent of the full scale signal of the channel input. This test checks all the harmonics depending on the sample rate. The DAS-1A has a built-in test oscillator that operates at 20 Hz and at a level of 20 mV.

Filters

The sigma–delta analogue-to-digital converter over-samples the data by 256 times at 1 ms, which gives an effective Nyquist frequency of 128 kHz. This means the analogue anti-alias filter can be very simple. This is implemented by three single R-C filters with cut-offs of 7 kHz (nominal) distributed through the amplifier signal path. This reduces the system noise without affecting the overall phase/amplitude response within the normal seismic passband. Anti-aliasing is done with a digital filter and decimation to produce the required sample rate (Table 2.5). The filter cut-off is fixed to the sample rate and is over 0.8 of the Nyquist frequency.

Table 2.5 Digital anti-alias filter (finite input responses)

Sample rate (ms)	Passband (Hz)	Passband ripple (dB)	Cut-off (−3 dB) (Hz)	Stopband (Hz)	Group delay (ms)	Decimation (×1)
$\frac{1}{4}$	1500	0.2	1625.5	2000	7.25	64
$\frac{1}{2}$	750	0.04	824.3	1000	14.5	128
1	375	0.08	411.9	500	29	256
2	187.5	0.1	205.9	250	58	512

The low frequency cut-off points are determined by plug-in resistor packs for both filter in and out settings. The filter settings are 8, 12, 16 and 20 Hz. The filter pack allows four discrete low frequency cut-off points. These are nominally the 3 dB attenuation points of the filter. The filter slope is fixed at 18 dB/octave. The filter out setting selects the lowest frequency value of the filter module, but reduces the slope to 6 dB/octave. This means the absolute low frequency response can be pre-programmed. The rest of the amplifier has a DC response. The system tests should check the frequency response of the system and confirm that the high cut and low cut filters are set correctly.

Gain accuracy test

In the OYO DAS-1A system the signal generator is used for this test. A set of 2048 samples is collected and the middle 2034 samples are used to find the true gain of individual channels. Channel-to-channel gain errors are calculated as percentages.

The Geometrics Strativisor NX recording system uses essentially the same test suite as TTS-2 and OYO DAS-1A. This can be summarised as follows:

- Instrument noise
- DC offset
- Crossfeed
- Common mode rejection
- Gain similarity
- Gain accuracy
- Phase similarity
- Filter tests
- Bandwidth
- Timing accuracy

2.8 Sparker seismic energy sources

For many years the best known and most well-tried seismic energy source for engineering site surveys was the 'sparker' system. In 1956, Knott and Hersey discovered that a capacitor bank can discharge its energy through an underwater electrode and the resultant spark produces an expanding bubble of ionised gases which constitute a seismic pulse capable of penetrating several hundred metres of strata below the seabed.[11] In 1957, Alpine Geophysical introduced the first commercial sparker seismic system. Most sparker systems have now been superseded by airguns of one type or another, though there is some residual interest in resurrecting sparkers as an alternative to airguns. Sparkers were, for example, used in the original Aquatronics high resolution profiling system. This gave a short pulse of 10 ms, with a peak frequency in the region of 100 Hz. The Aquatronics sparker was a 3.0 kJ system that proved inadequate in the northern North Sea. This was particularly the case where over-compacted glacial clays and outwash reached thicknesses of 150 m or more. In order to maintain the short pulse and frequency content, a 15 kJ ten-tip source was developed.[12]

The sparker system is literally an underwater spark. The principle of operation is relatively simple. An electrode rod insulated from but physically close to a ground frame is immersed in the sea at high potential of between 4000 and 15,000 V. This voltage is derived from an electrically charged capacitor bank and is suddenly connected between the electrode rod and the ground frame. A heavy surge of current begins to flow. Since the electrode rod is insulated and only a small end surface area is exposed to the conducting seawater, the current flow and potential field will be something like that shown in Figure 2.7. The potential gradients and current densities become very high at the end surface of the electrode rod. Since the end surface is to some extent rough and pitted, current flow will be even more strongly concentrated at sharp points and spines on the metal surface. Since the heat generation is concentrated where the electrode current density is greatest, the spark is in essence a flash boiler at the electrode surface. An extremely hot and incandescent plasma of steam, ionised gases and vaporised metal 'gas' and free electrons is formed. The expanding plasma bubble forms the sparker

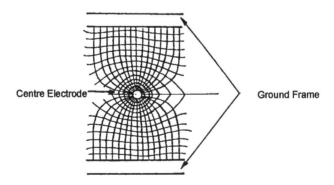

Figure 2.7 Sparker potential field diagram.

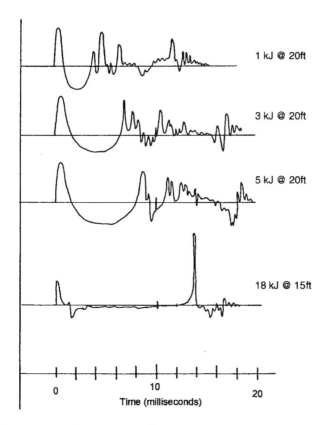

Figure 2.8 Sparker signatures diagram.

'flash', which may be observed during night-time operations. Sparker pressure signatures are shown in Figure 2.8.

The frequency spectrum produced is a factor of the power per tip of the spark array used. The higher the power per tip the lower the frequency spectrum and vice versa. For digital seismic surveys a power per tip of 0.75–3.0 kJ is usually used. This produces a frequency spectrum in the range 60–130 Hz. The main drawback to sparker discharge systems was always pulse control. The slightest variation in the physical geometry of the system and consequently in the electrical resistance of the discharge path caused large fluctuations in the seismic pulse and a resultant degrading of the seismic data.

Figure 2.9 shows a basic sparker schematic. The generator power supply is transformed from 220/240 V to the full charging voltage, typically 3500 V (Figure 2.10). The trigger units should have sufficient capacitors to provide the power output, anything from 1–15 kJ. In the schematic shown each 1 kJ trigger unit has a 3.5 kJ energy storage unit connected. This would give a total of 4.5 kJ to each set of three tips, 13.5 kJ in total. The sparker is fired by means of

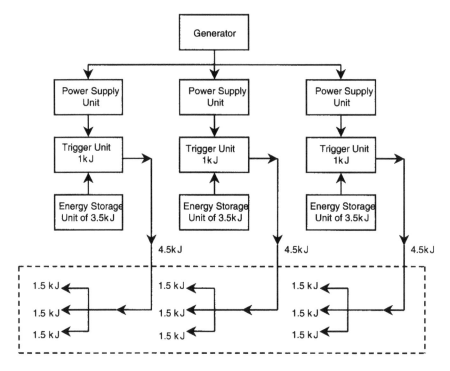

Figure 2.9 Schematic of 13.5 kJ digital sparker.

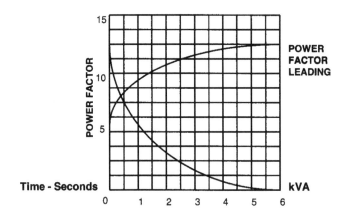

Figure 2.10 Power supply full-load power factor characteristics.

a triggered spark gap (Figure 2.11). The gap is too great for the spark to bridge the gap at 3500 V. However, the trigger from the fire control unit and trigger isolator operates at a much higher voltage and can ionise the gap and discharge the capacitors across the sled towed in the sea. The sled shown is a

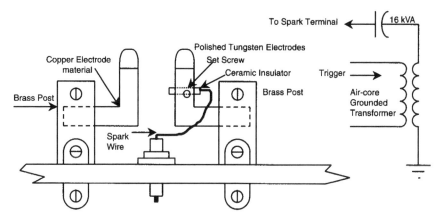

Figure 2.11 Sparker firing mechanism (vertical rail spark gap).

standard EG and G nine-tip sled. Each tip discharges typically 1.5 kJ of energy in the space of a few milliseconds. Since the energy stored is related to the square of the charging voltage the power output at different voltage levels can change quite radically. Thus for a 160 µF capacitor, a type used for years by site survey companies, the power output in watt-seconds can change from 925 J at 3.4 kV to 1160 J at 3.8 kV.[13] This is shown in Figure 2.12.

Most sparkers used for site surveys are the EG and G 230-series units, or variations on this design, and have been in service since the 1960s. The limiting factor on the EG and G sparkers is the power handling capacity of the trigger units and the power supply units. The maximum firing rate for this type of sparker is as follows:

$$\frac{\text{Total rated energy stored (kW/s)}}{\text{Number of power supply units used}} \times 1.6 \text{ s}$$

The method of triggering is shown in Figure 2.11. A very high voltage at 16 kV ionises the gap between the tungsten electrodes and the capacitive discharge takes place. A further limitation on the sparker operation is the phase relationship of the power factor which changes with time and generally speaking sparker firing rates below 0.5 s cannot easily be achieved. Figure 2.12 shows the power output at various voltage levels. This illustrates in a sharply focused way the importance of establishing the voltage output in relation to the sparker power output. Figure 2.13 shows a typical section of single electrode sparker record.

Quality control on sparker arrays

Quality control on sparker systems involves a number of system checks which may be listed as follows. First, an opto-isolator is necessary, connected

48 High resolution site surveys

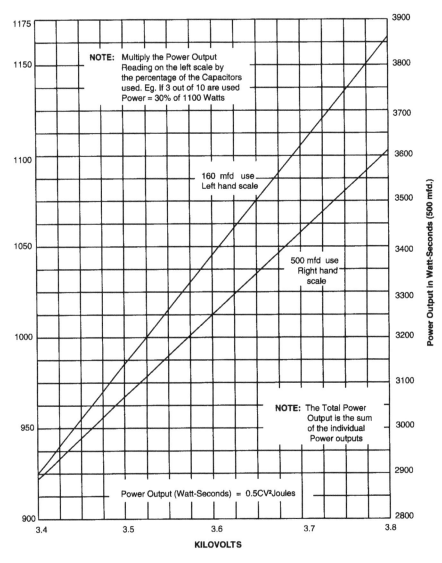

Figure 2.12 Sparker power output at different voltage levels. Sparker firing rate = 2000/1000 Ws = 2 pulses per second for one power unit. Total power (in watt seconds or joules) = $0.5\, CV^2$ where C is in microfarads and V is in volts with 160 µF capacitors and $V = 3.53$ kV. Total power = 0.5 (160) × (3.53) = 1000 J.

between the fire control unit and the digital field system (DFS). The opto-isolator prevents any spurious sparker-generated radio frequencies entering the fire control unit and the DFS V controller unit. The controller unit is especially vulnerable to damage by sparker-generated radio frequencies.

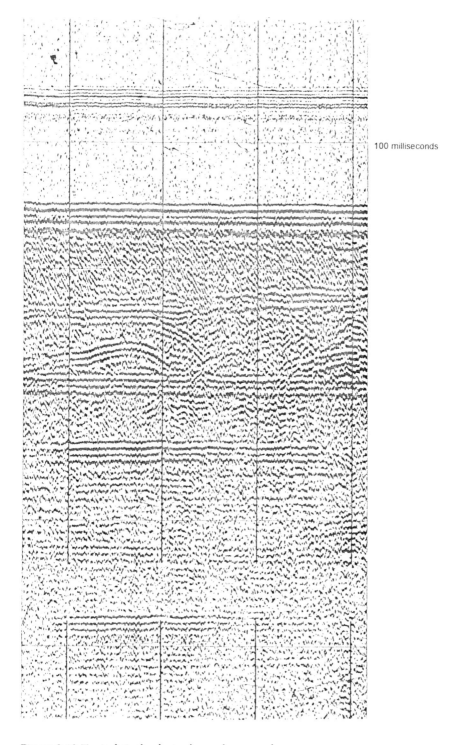

Figure 2.13 Typical single electrode sparker record.

> **Box 2.2**
> **A tale of an inexperienced seismic crew**
> Once upon a time a site survey was set up with a sparker seismic source and a brand new digital recording system. The crew did not know about the necessity of an opto-isolator circuit. Asked if they thought this was necessary the author was told to mind his own business. The first time the sparker fired the digital system was effectively wrecked by radio frequency pick-up from the sparker. It took a week to obtain new printed circuit boards. The crew did not know how to build an opto-isolator circuit and the author had to do this for them. This was not really the fault of the crew. The boom-and-bust nature of the oil industry ensures that nearly all seismic crews are inexperienced, since few people last more than one boom period and disappear in the next bust cycle.

The sparker electrode power must be very carefully checked. Increasing the power to an electrode tip reduces the frequency spectrum. It has already been noted that since the capacitor charging is determined by the square of the voltage, a small shortfall in the charging voltage can cause large changes in the sparker power output. Before the start of the survey the sparker generator should be checked. It should be of sufficient power to charge the capacitor energy storage units, at least 40 kVA for a 15 kJ sparker system. The generator should be dedicated to the sparker system and not used for instrument power supplies. The generator should be earthed with a single point welded earth. Earthing problems and earth loops are a major source of downtime when sparkers are used. The condition of the sparker tips is also critical. The tips should be flat and the insulation intact. Tips that have burnt back into the insulation give a degraded pulse due to slow current build up. Individual tips often burn differentially, altering the power of those that remain and altering the pulse.

The main reason that sparkers were abandoned as an energy source for high resolution surveys was the multiplicity of causes for pulse distortion. Slight geometry changes in the towed units, slight voltage charging deficiencies, problems with very large electrical generators, problems maintaining electrical insulation with in-sea cables and problems with radio-frequency-generated interference from the discharge units all mitigated against further development of the sparker system. It should also be remembered that these high voltage electrical systems were inherently dangerous to the operators and engineers.

2.9 High frequency seismic airgun sources

The main alternative to spark arrays is an airgun or perhaps a small airgun array. The use of small airguns as an alternative to spark arrays has been commonplace since the 1980s. Airguns are the most widely used non-explosive seismic energy source for all types of survey, but have tended to be superseded by water guns and sleeve guns. The airgun works on the principle of ejecting a high pressure (2000 psi) bubble of air into the water at a precisely known moment in time.

For engineering site surveys a single small airgun of perhaps 40 in.^{3} may be used. Alternatively an array of up to four small guns is a popular choice. The frequency spectrum of the system is likely to be in the region 40–120 Hz, comparable to the sparker system previously described. Figure 2.14 shows the operation of a typical airgun. High pressure air charges both the upper and lower chamber. The piston assembly is held in the downward position because the area of the trigger piston is larger than that of the firing piston. To fire the gun an electrical pulse opens the solenoid valve, lifting the piston assembly and equalising pressure above and below the trigger piston so that the compressed air in the lower chamber drives the piston assembly forcibly upwards, releasing the air in the lower chamber into the water. It is this explosion of compressed air which provides the primary seismic pulse.[14] Figure 2.15 shows the airgun pressure signatures at different depths.

The dominant frequency of a pulse generated by an airgun is controlled by the air pressure, the size of the lower chamber and the depth of operation. When the gun fires the bubble of air expands outwards and then collapses in on itself. The bubble oscillates and a seismic wave will be generated on each oscillation. For site survey work a single gun with a strong primary pulse is an advantage, since the bubble pulse oscillation actually degrades the seismic

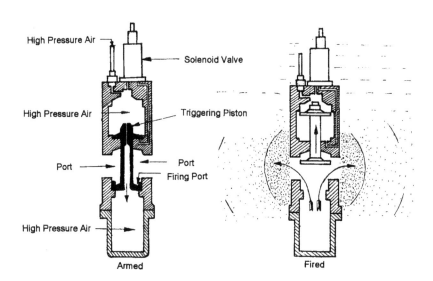

Figure 2.14 Typical site survey airgun. (Reproduced from Milton B. Dobrin, *Introduction to Geophysical Prospecting*, p. 124 by kind permission of McGraw-Hill Publishing Company.)

52 High resolution site surveys

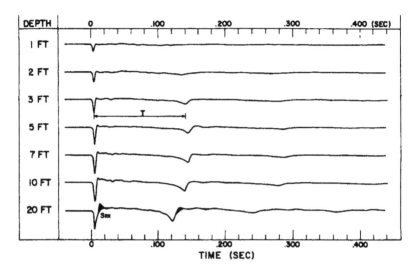

Figure 2.15 Airgun signatures at different depths. (Reproduced from Milton B. Dobrin, *Introduction to Geophysical Prospecting*, p. 125 by kind permission of McGraw-Hill Publishing Company.)

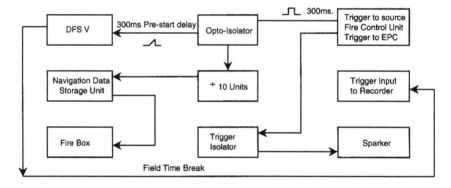

Figure 2.16 Typical airgun or sparker firing sequence.

record. A small array will have some bubble cancellation. Typically, three 40 in.³ airguns will provide the primary seismic pulse with a 20 in.³ gun providing a degree of bubble cancellation. Figure 2.16 shows a typical firing sequence schematic diagram for a sparker or airgun energy source.

Another, perhaps more modern type of airgun is the Seismic Systems Inc. GI airgun. Effectively, there are two independent airguns within the same casing. The first element is the generator which generates the primary pulse. The second element is called the injector as it injects air inside the bubble produced by the generator. It is used to control and reduce the oscillation of the bubble created by the generator. Each gun has its own reservoir, its own shuttle, its own set of exhaust ports and its own solenoid valve. A common hydrophone provides the time break and the shape of the near field signal. The gun phone is located inside the bubble and therefore it responds to the air blast of the GI gun to which it is attached. Figure 2.17 shows far field signatures of this gun system for a single 90 in.^{3} gun and a two-gun array at 300 in.^{3} (210 + 90 in.^{3}).[15]

When the generator is fired, the blast of compressed air produces the primary pulse and the bubble starts to expand. When the bubble approaches its maximum size, it encompasses the injector ports and its internal pressure is far below the outside hydrostatic pressure. At this time the injector is fired, injecting air into the expanding generator bubble. The volume of air released by the injector increases the internal pressure of the bubble and prevents its internal collapse. The oscillations of the bubble and the resulting secondary pressure pulses are reduced and reshaped. The gun signature can be reshaped at will by adjusting the volume of the generator and/or the injector. The time of injection also reshapes the gun signature. The duration of the injection can also be altered by means of exhaust port reducers.

Quality control for airgun sources

To establish a competent level of quality control on airgun seismic sources nine basic topics need to be addressed. These can be discussed as follows.[16]

Far field tests

To perform a far field test the gun must be considered a point source in the water. A hydrophone is placed underneath the gun or perhaps an array of guns, usually at a depth of ten times the length of the array. For an array length of 15 m the hydrophone needs to be at a depth of 150 m. To avoid the seismic pulse being distorted by the proximity of the seabed, the total water depth should be three to four times the hydrophone depth. The hydrophone should be positioned exactly under the array by firing the near and far guns and looking at the direct arrival from the guns on the oscilloscope. When the hydrophone is in position the array is fired and the resultant signature can be stored on magnetic tape for computer analysis. The effect of individual guns can then be analysed and the allowable drop-outs determined. These remarks may appear to apply to a 2D seismic array but sooner or later any source used for survey work needs to be evaluated by far field signatures. Computer modelling of arrays and gun combinations is accepted by most oil companies but the requirement for far field tests can never be completely negated.

Time break generation

If all the guns on an array are fired in synchronisation the interval between the initial pulse and the first bubble pulse will be different for each gun having a

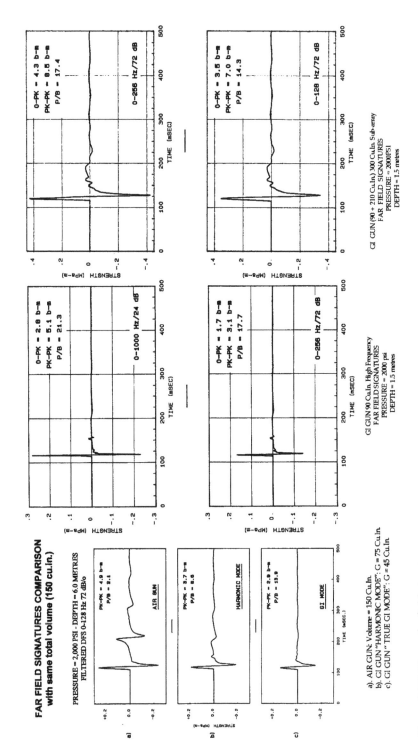

Figure 2.17 GI gun far field signatures.

High resolution digital site survey systems 55

different air capacity. Obviously the different bubble pulse periods can be made to mutually cancel one another but this only happens if the array firing is very accurately timed. This means that the firing sequence of the guns and the array timing must be carefully checked. From the quality control viewpoint, establishing the time break sequence and the timing of the array is very important.

Near field signature

The near field signature can be observed from a hydrophone attached to the array itself. The signature will be corrupted by the proximity of the airguns and will not be a true visual relationship. The advantage of a near field signature is that it can be monitored during data acquisition. If the signature is noted after timing any change in timing can be observed while on-line. Some idea of pulse accuracy can also be made and this is a useful adjunct to quality control.

Air pressure

Air pressure should be checked, usually 2000 psi ± 10 per cent. Air pressure should be checked on the main air manifold as well as on a gauge in the instrument room.

Timing

Timing should be checked for each shot on every line. Autofiring (the random firing of the gun) should be checked for periodically by using an oscilloscope on continuous sweep or by playing back records in the fixed gain mode on the digital recording system. AGC mode tends to mask small gun autofires. Any autofire on a seismic record automatically ruins that record.

Source power

The source output and flat frequency response should be checked in processing. There are three elements to be borne in mind when considering power output: first, the peak pressure; second, the front-to-back ratio, that is, the peak pulse to bubble pulse ratio of up to 15:1; finally, the frequency spectrum can be measured by looking at the pulse.

Depth control

When an airgun fires in the water the period of the bubble pulse oscillation varies with depth, so accurate depth control is important. Typically a four-gun site survey array will be towed at a depth of six feet. Site survey sources are usually buoyed but a manometer type of depth indication should be considered necessary for modern site surveys.

Electrical and mechanical condition of the array

The physical state of the airguns must be checked. The mechanical condition of a particular airgun can be tested by pressuring the airgun to its design pressure. By shutting off the line valve any pressure drops can be observed. These could be caused by worn O-ring seals, bad lines or connectors, bad solenoid seatings, etc. Additionally, at the start of survey the level of spares carried should be known. The electrical leakage, that is, the tendency to a short circuit condition on solenoid and sensor lines, must be checked as leakage can cause unwanted firing delays.

Compressor capacity

Before the start of survey the airgun compressor capacity needs to be established, together with its capability to maintain the array at the contractually stated air pressure. For a 2000 psi array a variation of ±10 per cent is usual. The power output of a gun array should also be established before the start of survey. Site survey energy sources usually operate at 5–15 bar m. The contractor should be asked to provide evidence of the peak pressure, the front-to-back ratio and the frequency spectrum.

As an actual example of what can go wrong, let us suppose that air is supplied to a 150 in.^{3} source array by two compressors each capable of producing 75 ft^3 of air per minute. The source system can be matched to the compressors by the following calculation:

Box 2.3
The gap between geophysical managers and field crews

In the case quoted here both the oil company geophysicist and the survey company geophysical salesperson assumed that an airgun array of 150 in.^{3} could be run from two compressors supplying a combined air capacity of 150 ft^3/min. How they managed to confuse the cubic capacity of the source array with the compressor capacity and assume that if the two were equal the array could be operated is unfathomable. This is manifestly not the case as the calculations shown here indicate. In fact, neither manager had the necessary field crew experience to work out how the compressors are matched to the source in use, an unsurprisingly ordinary occurrence. The survey company then blamed their party manager onboard the survey vessel, while the oil company blamed their consulting engineer, again onboard the survey vessel. Yes, gentle reader, the author of this book got it in the neck yet again.

High resolution digital site survey systems 57

- 150 in.³ source array: $0.0868\,\text{ft}^3\{150 \div (12 \times 12 \times 12)\}$
- Air pressure requirement: 2000 psi (135 atm)
- Air per shot: $11.718\,\text{ft}^3$ of air per shot, i.e. 0.0868×135
- Shots per minute: 20 (3 s firing for 6.25 m shotpoint interval)
- Air requirement: $234.36\,\text{ft}^3$ per minute (cfm), i.e. 11.718×20
- Air supplied: 150 cfm

Fairly obviously the contractor could not meet this air requirement, since the compressors available produced only 150 cfm. Two solutions presented themselves, first to increase the shotpoint interval to 12.5 m and accept 50 per cent fold. That is, with a streamer group interval of 12.5 m and a shotpoint interval of 12.5 m the coverage would be halved. The second solution was the one adopted, which reduced the source array to a single 90 in.³ gun. Again the compressors can be matched to the source in the following manner:

- 90 in.³ gun: $0.052083\,\text{ft}^3\{90 \div (12 \times 12 \times 12)\}$
- Air requirement: 2000 psi (135 atm)
- Air per shot: $7.03125\,\text{ft}^3$ of air per shot, i.e. 0.052083×135
- Shots per minute: 20 (3 s firing interval)
- Air requirement: $140\,\text{ft}^3$ per minute (cfm), i.e. 7.03125×20
- Air supplied: 150 cfm

In this case the compressors were capable of supplying the source array, keeping a shotpoint interval of 6.25 m and achieving full (100 per cent) coverage.

2.10 Other site survey seismic energy sources

Sleeve guns and water guns are also used by some contractors. An advantage of a single sleeve gun or water gun is the near absence of a bubble pulse which obviates the need for any bubble cancellation. In the case of sleeve guns the storage chamber is designed to open more rapidly than is the case with a more conventional airgun. This increases the rate of air discharge and the peak pressure of the output signal, with most of the energy concentrated in the primary pulse.[17] Water guns are sometimes used for site survey work. Often they are modified airguns with one chamber filled with water rather than air. Water guns produce their acoustic output from the sudden collapse of a cavitation volume in the water.[18]

Other older systems include the Aquatronics 'closed frame' sparkers, the Teledyne 'snake' opposed tip sparkers, the Exxon mini-sleeve exploder and the Geomechanique Flexichoc system. Some of these systems remain in existence but it is some years since they have been seen in an operational role.

2.11 Source fire control systems for site surveys

An important element in quality control on sparkers and small gun arrays is to establish the firing sequence. A fire command will usually be derived from the navigation system, almost inevitably differential mode GPS processed by an integrated navigation system. This 'time zero' is fed to the energy source control unit which gives two outputs, one to the digital recording system and one to the energy source timing unit itself. The digital recording system then writes the header information to the data tape. After a delay of typically 64 ms the digital recording system generates a 'start of data' signal, often known as a wire blast. The wire blast is the remote start from the digital recording system to the energy source timing unit. The energy source timing unit needs two timing commands to fire the array, the navigation start command and the wire blast. On receipt of the wire blast the seismic source is fired. There are other methods of generating a time break but most systems in use have two inputs as a requirement for firing the guns and starting the digital recording system.

Occasionally a source synchroniser is used that is more usually found on a 2D seismic survey vessel. The most common such unit is the Litton Resources systems energy source synchroniser (LRS-100), generally considered obsolete for 2D work. Figure 2.18 shows a schematic of the LRS-100.[19] Figure 2.19 shows the various fire commands, responses and their relationship in time. This unit is a microprocessor-based system and controls the firing of all guns in the array. The gun control module receives, upon firing, a return signal from the gun transducers located in the air supply line, which indicates a 'fire detect' signal. This fire detect signal nominally occurs at 128 ms from the start of the fire pulse to the gun solenoids. If the fire detect signal does not occur at 128 ms the error is corrected automatically at the next fire command. After firing the data is sent to the display controller module (DCM). The DCM receives the firing and control data and displays it on a video monitor. Data is also printed out as hard copy.

Having established the time break and timing sequence the near field signature can be established using a hydrophone on the array itself. The signature will be corrupted by the proximity of the individual guns or sparkers and will not be a true visual relationship; however, the advantage of a near field hydrophone is that it can be monitored during shooting. If the signature is noted after timing, possibly on a Polaroid camera, any changes in timing can be observed while on line. The use of a near field signature is therefore a useful adjunct to quality control.

After the gun array has fired, a sensor on each gun inputs a signal to the source timing unit and the exact time each gun fired is recorded. Corrections can be applied and each gun corrected to fire at the same instant. Should the automatic timing fail, each gun can be manually timed on a storage oscilloscope. Timing a small array of perhaps four guns on an oscilloscope is relatively common for site surveys.

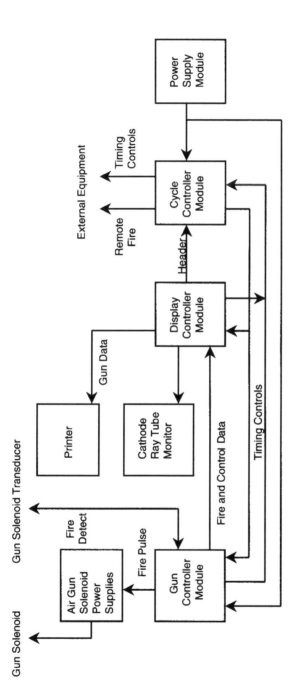

Figure 2.18 Seismic energy source control and firing system.

60 High resolution site surveys

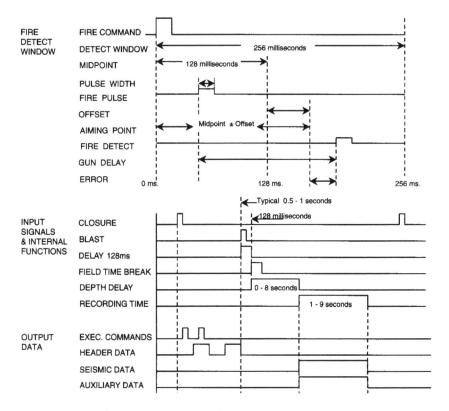

Figure 2.19 Fire detect, input signals and output data.

Sparker energy sources and single airgun sources tend to use a simplified form of time break generation. Typically such systems will be fired on a time basis, derived from vessel speed. A system record annotator is programmed to give the fire command at a manually set fixed interval. The requirement is that the sparker or airgun fires at a shotpoint interval of say 6.25 m. This distance will be recorded by the navigation data logging system. The record annotator gives a system start to the DFS V. There is then a delay of typically 300 ms, while the header data is written. After the header has been written a blast command goes to the sparker or airgun. When a sparker fires, a transformer on the system output gives the sparker fire time to the DFS. The DFS starts recording when the blast command is issued but records nothing but zeroes until the sparker firing pulse is inputted back to the DFS.

2.12 Streamers for site surveys

Streamers used for site surveys are usually Teledyne or Fjord analogue seismic streamers used as twenty-four, forty-eight or ninety-six trace streamers.

The Geco-Fjord digital streamer is offered by at least one contractor. The streamer group interval can be 6.25, 12.5 or 25 m in length, though most contractors, as previously stated, use a fixed 12.5 m group interval. Most streamers used for site survey work use linear arrays of hydrophones, perhaps eighteen hydrophones per 12.5 m group. It is by no means axiomatic that a linear array of hydrophones should always be used and logarithmic arrays have been used in the past. Most site survey streamers use transformer coupling so the sensitivity of the groups is not variable. In fact, there is no reason why a charge-coupled system should not be used and it is believed that one contractor uses or has used this variation. A modern development is the GCCC seismic mini-streamer for site surveys. This is a streamer designed specifically for site survey work and is a small diameter streamer (1.9 in.), split into 100 m sections. Figure 2.4 has already been referred to and shows a full system schematic with the streamer connected to the DFS V system.

The depth at which streamers and seismic sources should be towed is dependent on a number of factors. When towed near the sea surface the secondary signals reflected from the sea surface can cause destructive interference with undesirable notches occurring at frequencies dependent on the tow depths. This effect is known as ghosting, and if the notches occur within the desired seismic bandwidth the useable bandwidth will be reduced and the resolution degraded. The seismic source and streamer should be towed at a depth equivalent to one quarter of the wavelength of the desired frequency. Table 2.6 gives the optimum tow depth for various frequencies.[20]

Streamer depth control

As with all other streamers used for seismic work, site survey streamers are ballasted to be neutrally buoyant. This minimises noise caused by the flow of water across the streamer skin. The streamer sections are filled with deodorised kerosene and the buoyancy is altered by varying the amount of liquid pumped into the sleeve. Sufficient oil must be present to maintain the cylindrical shape of the cable and prevent cavitation of the bulkheads. On a

Table 2.6 Optimum tow depth for various fundamental frequencies

Frequency (Hz)	Optimum tow depth (m)
50	7.5
100	3.8
200	1.9
500	0.8
750	0.5
1000	0.4
2000	0.2

62 High resolution site surveys

typical 300 or 600 m streamer three to six depth transducers will be used. Most depth indicators convert the output of a variable reluctance transducer to a DC signal which is proportional to depth. The signal is corrected logarithmically for any transducer non-linearity and displayed.

Streamer depth control is normally maintained by Syntron or Digicourse depth controllers. The principle of operation of depth control birds is that of two forces acting on a piston diaphragm assembly. The cable depth required is set on the birds prior to cable deployment. A spring is compressed to a point where at the required water depth the compressed spring is exactly balanced by the static water pressure so that the diaphragm is in a neutral state. When the force exerted by the spring is greater than the force exerted by the static water pressure, the diving planes assume the dive position. The depth controllers used for site survey streamer operations are the same as those used for 2D/3D seismic surveys. Perhaps the most typical depth control unit in use today is the Digicourse 5010 Digibird.[21]

Streamer noise

Noise checks form a large part of assessing a hydrophone streamer. Often exact noise and sensitivity specifications are not stated for a particular contract, it being left to the discretion of the consulting engineer to decide what is reasonable. For most systems in use a sensitivity of 5 µV per microbar is typical. A sometimes contractually stated ambient noise level of 3–5 µbar on the streamer traces and 5–8 µbar for bird (depth control) and near traces is typical for a Teledyne streamer. The cable should initially be deployed without depth controllers to establish a basic trim, otherwise the depth control birds will generate excessive noise when attached. The depth detectors should be calibrated at the sea surface and at a towing depth. Preferably this should be done before the start of survey. Depth control birds should be checked for correct pressuring and for free movement of the shoes. Depth control birds should be located near group centres. Inter-channel crossfeed between the different channels should be checked on a daily basis. Neither the streamer tailbuoy nor the vessel should generate cable jerk. If any significant cable jerk is present then the stretch sections, tailbuoy or towing configuration are suspect.

Figure 2.20 shows a section of sparker record illustrating the different degrading effects that noise and incorrect streamer depth control can have on a seismic record. There is noise caused by the streamer being in a turn and of changes in pulse width due to fluctuations in streamer depth, A–B on the record section shown. The section of record B–C shows a good data record with the streamer at a correct depth of 12–15 ft. Finally, the record section C–D shows how much the seismic pulse width changes with deviations in streamer depth. The importance of good streamer depth control cannot be over-emphasised. For high resolution seismic work it is usually best to tow the seismic source and the hydrophone array (streamer) within one

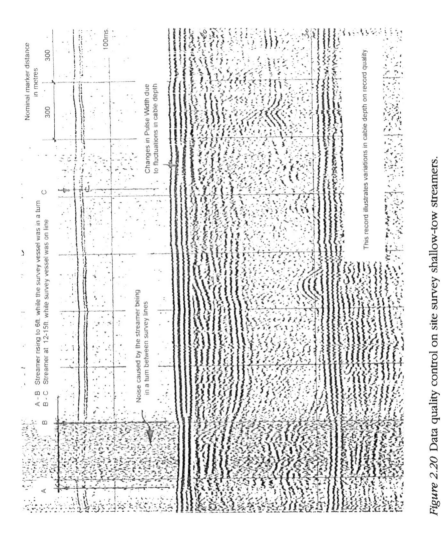

Figure 2.20 Data quality control on site survey shallow-tow streamers.

64 *High resolution site surveys*

half-wavelength below the sea surface. The wavelength in question is the dominant frequency in use. For typical high resolution site surveys with a frequency spectrum of 80–150 Hz a tow depth of 10–12 ft may be taken as standard.

Streamer leakage and continuity

As with any conventional seismic streamer, leakage and continuity form an important part of any system checks on a streamer. For site survey work streamer leakage is often measured on the DFS V analogue modules. Kalamos cable tester units are preferable and give more information about streamer faults. Most contracts state an allowable level for streamer leakage, typically 2 MΩ per channel with a downward limit of 0.5 MΩ. Cable leakage should be checked daily or whenever the cable is recovered for repair. The 2 MΩ limit should only be relaxed in special circumstances.

Streamer polarity

Streamer polarity should obey the standard seismic convention that a positive pressure displacement produces a downbreak on a camera galvo and a negative number on tape. Polarity can be established in a number of ways. Each streamer section can be tap tested on deployment. A small airgun can be fired on deck at low pressure. The twelve sections nearest the surface of the reel can then be checked for the downward sweep of the camera galvoes. As successive groups of twelve traces are exposed they can be checked.

Streamer sensitivity

Streamer sensitivity can be checked using a calibrated hydrophone. Sensitivity can be defined as the acousto-electric transduction factor of the array of piezoelectric sensors mounted in the streamer sections, as determined for acoustic wavelengths which are large compared to the array dimension. Sensitivity is normally expressed in volts per bar or as microvolts per microbar. Values in the range 2–50 V/bar are normally encountered in the site survey industry. As long as the combination of streamer section and seismic instrumentation does not significantly affect the signal-to-noise ratio, that is, if the recorded noise is mainly of acoustic origin, then the absolute value of the sensitivity is in itself not very important.[22]

Calibration serves two purposes: it checks the manufacturer's specifications and it identifies recorded electrical noise in terms of streamer-generated acoustic noise. A section of streamer is coiled and hung with the calibrated hydrophone inside the coiled section, horizontally, at the towing depth. The streamer section must be coiled to suppress the directional effect of the streamer array section. The coil should represent as small a diameter as possible, so that the diameter dimension is much smaller than the relevant acoustic wavelengths used in the test. For example, the acoustic wavelength of a 50 Hz signal will be about 30 m so the coiled streamer section has a 30 : 1 difference ratio. A single shot is fired and the amplitude of the signal, as viewed on a monitor camera through several different filter settings, can be assessed. The primary peak seen on the monitor camera or oscilloscope should not be used because the primary peaks are affected by bandpass differences. The first bubble amplitude should always be used for calibration purposes. The formula for calculating the streamer section sensitivity is as follows:

Streamer sensitivity
$$= \frac{\text{Streamer section amplitude}}{\text{Calibrated hydrophone amplitude}} \times \text{Calibrated hydrophone sensitivity}$$

The calibration process can then be repeated using the second bubble oscillation which should show a good correspondence with the first bubble calculation.

Streamer feather

Streamer feather angle, that is, the angle the towed hydrophone array (streamer) makes in relation to the designated survey line, is often considered relatively unimportant on site surveys, but it should be pointed out that excessive feathering will degrade lateral resolution when the data is CMP (common mid-point) stacked. This effectively increases the size of the area from which reflections are received. It is certainly the case that excessive feather angles are not usually seen on short streamers but a means of tailbuoy tracking should be considered necessary even if it is only a radar reflector attached to the tailbuoy. In the main feather angles over 10° should not be allowed. Most site survey contracts state an allowable feather angle of 6°–10°.

Streamer compass data

Cable compass data can be obtained from systems such as Digicourse Digibirds. These units are more usually associated with 3D surveys but compass birds are increasingly used on site surveys. Typically, three units might be used to give a basic streamer shape. Digibirds are often combined compass and depth birds. The model 395 cable depth controller is a microprocessor-based system mounted externally on the streamer. The unit is streamlined, battery powered and compatible with existing mounts and communication coils. Communications with the bird occur over a single twisted pair transmission line, using inductive couplings operating on a frequency of 27 kHz. The Digibird 396 compass operates on an optical scanning process using a compass element suspended in a pool of mercury. The data processing is controlled by a 1024-bit digital processor. It uses a twisted pair transmission line. The local magnetic field strength is sensed and translated into co-ordinates that give a magnetic heading. This data is accurate through a roll at ±45° pitch. Accuracy is ±0.5° relative to magnetic north.

> **Box 2.4**
> **The truth about measuring streamer feather angles**
> Once upon a time there was a seismic survey operation with a 1200 m streamer. A streamer tailbuoy was attached to the end of the streamer. The tailbuoy had a radar reflector attached and the feather angle was measured by taking a radar bearing to the radar reflector. Since the streamer was fairly short the radar reflector tended to be lost in the 'clutter' near the centre of the radar display. Whenever the streamer feather angle was measured, the survey party manager could somehow always see it through the radar clutter and somehow the feather angle was always inside the contract specification. Whenever the oil company consulting engineer looked at the radar display somehow the radar reflector seemed lost in the radar clutter and it was suggested that perhaps the tailbuoy was no longer attached. 'What nonsense', exclaimed the party manager, 'I can see the tailbuoy perfectly clearly'. At that moment the captain nudged the author and pointed out of the bridge window. There, not 30 yards away was the streamer tailbuoy, complete with radar reflector, obviously detached and floating serenely by. To add insult to injury the name of the survey vessel was written in large letters on the tailbuoy. Collapse of stout party, as they used to say in the back numbers of *Punch* magazine. Much the best modern method of measuring streamer feather angles is to have three compass birds in the streamer and to measure the feather angle from the compass values.

The resolution is 0.1°. The Digibird data is processed by a binary control unit. Compass data can be averaged over an operator-selected time and stored in a random access memory. The operator unit is a 16-bit minicomputer with a Zenith disc operating system. It contains 192 kilobytes of internal memory. A modem board facilitates communication down the transmission line to the birds. A common problem with Digicourse compass birds is that of calibration changes when the units are brought onboard during a survey. Significant calibration changes can sometimes be detected which reduce back to the original calibration values when the Digibirds are replaced and run for ten lines or so. Exposure to local ferrous metals onboard may account for this, with the time delaying residual magnetism accounting for the reversion to the original calibration values.[23]

2.13 Onboard processing

The advantage of an onboard processing system is very great indeed for site surveys. The surveys are often of short duration, and the results of the digital

High resolution digital site survey systems 67

system's test tapes processed on shore may only be available late in the survey or even after the survey has been completed. A rig move may follow very soon after the completion of the site survey and an initial assessment of the digital seismic data can be invaluable, particularly if shallow gas is present. Taking these factors into account there is an overwhelming advantage that accrues from use of onboard processing. Quality control on digital seismic data becomes objective and quantitative rather than subjective and qualitative.

There are a number of onboard processing systems available and the best known are Micromax and Promax. Less well known are systems such as Ramasses and Geoconvecteur. Micromax and Promax systems are well suited for quality control on site surveys but contractors are now offering a range of processing systems, often with their own in-house software. In terms of data input/output a large range of operating parameters is possible and a considerable array of geometry and headers can be inputted. Editing, muting and noise are standard features and any combination of trace editing is possible for quality control purposes. Amplitude/avo, statics, deconvolution, filtering transforms, data enhancement, velocity analysis, stacking and migration/DMO/depth conversion and wavelet processing are all often available onboard. The discussion that follows, covering processing techniques and parameters, is heavily reliant on the UKOOA guidelines for the conduct of rig site surveys.[24]

Quality control

In general terms, processing should be able to provide the following quality control checks which are UKOOA recommended but it may not be possible to produce all these parameters onboard the survey vessel. Figure 2.21 shows a section of on-line brute stack used for QC purposes (courtesy of Geometrics).

- Display near streamer groups for each shot on every line. This will check for timing errors and any source problems.
- Display at least one shot with all traces every fifty shots on every line. This will check for dead, noisy or reversed polarity traces.
- A brute stack section using a single velocity function should be produced. This will check general data quality.
- A velocity analysis should be produced for each line. This will ensure that the correct velocity trend has been picked.
- Iso-velocity displays should be produced for each line. This will check for velocity anomalies.
- NMO (normal moveout) corrected gathers should be displayed at each velocity analysis location. This will check that velocity corrections have been properly applied.
- Display each line with pre-stack processing applied. This is an alternative type of velocity check, as well as being a tool for the QC of pre-stack processing.

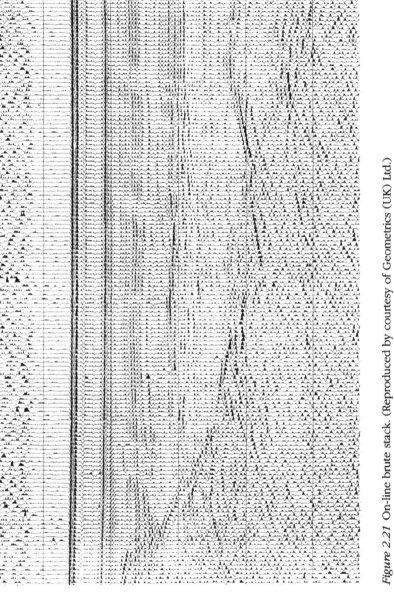

Figure 2.21 On-line brute stack. (Reproduced by courtesy of Geometrics (UK) Ltd.)

High resolution digital site survey systems 69

- Display the migrated data for each line. This will check that an acceptable output from migration has been achieved.
- Electrostatic display of the final processed data in scaled and relative form should be produced for each line.

Parameter testing

As a minimum requirement the following processes should be used for parameter testing:

- Designature
- Gain recovery
- Deconvolution before stack (DBS)
- Mute (outer and inner)
- Deconvolution after stack (DAS)
- Time variant filter
- Balance scaling
- Display trials (bias, polarity, dual polarity etc.)

Other processes might include demultiple, dip moveout, stack parameter tests, zero phase conversion, time migration and post-stack noise attenuation, that is, frequency versus wave number F-K dip filtering.

Designature

Most seismic sources have a far field signature recorded on an auxiliary channel. If this is not the case then a library source signature should be used. In extreme cases the seabed return can be used. The designature operator converts the far field source signature to its minimum phase equivalent. Designature should improve the vertical resolution by pulse compression. Most modern seismic sources give minimum phase and are very repeatable over long periods of time. In spite of this, designature should be applied to all water gun data.

Gain recovery and amplitude manipulation

It is very important to apply corrections for spherical spreading and absorption because a great deal of the interpretation is based on reflection amplitudes. It is possible to create or hide an amplitude anomaly merely by careless amplitude manipulation. Preserving relative amplitude relationships should be borne in mind throughout the processing sequence. Gain recovery used to be the first step in processing, using either a linear gain or a spherical spreading compensation function. The spreading compensation function is more usual for site surveys because it provides results which are more consistent than 'hand-picked' gain recovery functions. On contemporary surveys, gain recovery is usually applied at the post-stack sequence, though pre-stack processes such as deconvolution benefit from the input trace amplitudes being balanced beforehand. A simple exponential gain function is usually adequate. AGC (automatic gain control) is usually only applied before producing an equalised or scaled display. In the case of noisy data, long gate AGC is sometimes applied in the pre-stack sequence. This will attenuate high noise amplitudes but real reflection amplitudes will be adversely affected.

Partial pre-stack migration or dip moveout (DMO)

Dip moveout is sometimes used in heavily faulted areas, typically over diapirs, where conflicting dips can be resolved. DMO also improves cross-line ties, improves the attenuation of steeply dipping coherent noise and gives an enhanced velocity analysis. DMO can be costly and time consuming.

Velocity analysis

An accurate velocity model is needed before migration of data can take place. This is particularly the case with high frequencies. Incorrect velocity analysis does not merely degrade the data; seismic reflections can actually disappear. Time–depth conversion is usually based on RMS stacking velocities and the intervals at which velocity analyses are performed is critical. An ideal velocity sampling interval would be 50 m but intervals up to 500 m are common.

Deconvolution before stack (DBS)

This programme is applied to attenuate unwanted source signature effects, reverberations and multiples. An offset dependent DBS should be applied.

Mute

Outer trace mutes are needed to preserve vertical resolution because signal stretch on the far offset will degrade vertical resolution if allowed to enter the common mid-point (CMP) stack. Data on traces beyond a certain offset should be omitted from the stack by muting. Inner trace mutes are useful for multiple suppression, but care must be taken not to remove traces that contribute greatly to vertical resolution. Inner trace mutes should be applied to as few traces as possible.

CMP stack

The CMP stack improves the signal-to-noise ratio of the data more than any other processing step. There are a number of algorithms in use for factors such as normalisation, scaling and mute compensation. Weighted and median stacks are also relatively common. Offset weighted stacks should not generally be used for site surveys. Offset weighted stacks are good at stacking out multiples but degrade vertical resolution. Vertical resolution is always important when considering site surveys. If weighted stacks are to be used the effect of deconvolution after stack and FK demultiple should be assessed and a comparison between normal and weighted stack sections made.

Deconvolution after stack (DAS)

Predictive deconvolution after stack has the effect of removing seabed and multiple effects. The general aim is to improve vertical resolution by compressing the signal wavelet. Deconvolution is a process which works on the assumption that the signal wavelet in the data is of minimum phase. This means it should be applied prior to zero phasing. Deconvolution operator length is often selected at 1.3 times the water depth in milliseconds. Multiple removal then operates because the design window will be approximately ten times the operator length. This assumption tends to break down in deep water with short record lengths.

Time migration

Migration improves lateral resolution. It does this by either collapsing diffractions to the zero offset origin or collapsing the Fresnel zones. Migration benefits plane layered data as well as dipping data. There are numerous migration algorithms available and the choice of algorithm usually depends on the effect of spatial variation on the velocity field, while geological dip should also be considered. A good velocity model is absolutely essential and it should be borne in mind that FK migration techniques are bad at handling changing velocity fields. FK migration techniques should be avoided if there are lateral velocity changes caused by gas anomalies, channels, or major structure changes. Most migration algorithms are dip limited but this does not usually apply to site surveys, unless channel flank reflections or reflections overlying shallow diapirs exceed the dip limitations. Kirchhoff-based algorithms can handle steep dips but not laterally varying velocities. Kirchhoff migration is usually used where dip-limited migration algorithms cannot handle steep dips.

Post-stack coherent noise attenuation (FK dip filter)

The dip filter process removes unwanted coherent noise trains from the record section. The process applies a cross trace filter which can adversely effect lateral resolution. Dip limits are chosen with regard to geological structures and diffractions. Dip filter design should not be frequency limited and this type of dip filtering should not be performed before migration.

Zero phasing

The zero phase signal wavelet has the shortest duration and largest amplitude for a given amplitude spectrum. The zero phase wavelet helps distinguish an event against background noise and enhances resolution. Polarity reversals of seismic reflections are also more easily seen. Zero phasing can be achieved by a statistical means using multi-window design. Alternatively an uncontaminated wavelet can be taken from the data and converted to a zero phase equivalent. Zero phasing helps event picking algorithms follow the amplitude peaks. In turn, this allows extraction of the peak reflection amplitudes and the building of amplitude maps.

Time variant filter

For optimum resolution the signal-to-noise ratio should be maximised. Time variant filtering attenuates unwanted frequencies across the record section and helps improve the signal-to-noise ratio. Time variant filtering is a zero phase process and should be applied after zero phasing.

Balance and scaling

As a final step to producing an equalised, scaled record section, a sliding gate automatic gain control (AGC) is applied to the data. An AGC gate length in the order of 100–250 ms is normally used.

Data presentation

Survey data is usually limited to black and white, single polarity paper sections. This is starting to change and many different parameters and methods are now available. Plot scales can be varied considerably with compressed horizontal scales useful for determining structure. Expanded vertical displays can be useful for a detailed interpretation of a particular interval. Vertical scales at 20 or 30 cm/s are normal. Relative amplitude displays are plotted with a low gain level, typically 12 dB down on normal, and help interpret lateral amplitude variations. Equalised displays are plotted with the AGC applied at the standard amplitude level. Colour coding in relation to amplitude strength and instantaneous amplitude irrespective of phase are useful displays. Instantaneous phase displays, often referred to as continuity plots, can also be used. Amplitude versus offset (AVO) displays are particularly useful when dealing with gas-saturated sands. Data tape output is usually in the form of SEG Y CMP stacked data.

During data acquisition, as a basic minimum, at least a portion of each digital line should be processed. At least one wellhead location line should be processed in its entirety and an initial onboard assessment made of the possibility of shallow gas on location. A typical processing suite might be as follows:

1. Input seismic records from tape (SEG Y, 4 byte floating point)
2. Data resampled at 4 ms, only every second trace and second record
3. Semblance velocity analysis

High resolution digital site survey systems 73

4. Normal moveout (NMO) correction
5. Front end mute
6. Deconvolution if considered necessary onboard
7. FKKF coherent noise filtering (if necessary)
8. Time variant bandpass filtering
9. True amplitude recovery and display (6 dB/s)
10. CDP Stack (twenty-four-fold typical)
11. Plot

Further filtering and deconvolution can be applied as necessary. Basic interpretation can be performed onboard.

2.14 Digital site survey interpretation

For on-site digital seismic interpretation there are three fundamental aspects to consider. These are time–depth conversion, shallow gas detection and workstation interpretation.

Depth conversion is taken from velocity analysis during onboard processing. A mean set of interval velocities is determined and a scale bar drawn showing average depth in metres against two-way travel time. The scale bar is then used to measure section depths directly. However, since the conversion is based on RMS velocities, some care is necessary, especially in areas of high dip or complex geology. Given reasonably simple geology and careful velocity analysis, depths should be within about 5 per cent of their true value. Alternatively, where detailed borehole control is available, reflector depths may be calibrated directly.

Bright spots

Shallow gas detection is normally made, in the first instance, by recognition of anomalously high amplitude, usually phase-reversed reflectors (bright spots), allied to a number of other characteristics, such as acoustic masking, velocity pulldown, structural closure, edge effects, frequency reduction and basal flat spots. Gasified sands have seismic velocities that are typically in the region 1375 m/s, while water or oil-saturated sands have velocities that are in the region 2370 m/s. The lower velocity area of a seismic record produces a heavier lineation than areas where higher velocities produce a lower reflectivity. The high amplitude of the reflection area is usually known as a 'bright spot'. Figure 2.22 shows a record processed by conventional methods designed to equalise amplitudes. Reflections from the gas zone do not stand out. The same record is shown in Figure 2.23 processed to preserve relative amplitudes. The bright spot at 1.25 s stands out prominently delineated on a record by correct velocity analysis.[25] High amplitude phase-reversed reflectors are particularly relevant to shallow gasified sands, usually within the first kilometre below the sea floor. It should be emphasised that there will always be

74 *High resolution site surveys*

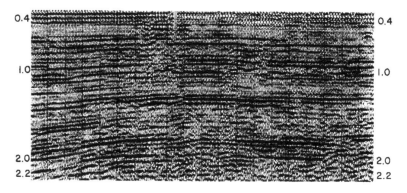

Figure 2.22 Conventional seismic record. Conventional processing is designed to equalise amplitudes. Gas zone reflections do not stand out. (Record reproduced from Milton and Dobrin, *Introduction to Geophysical Prospecting*, p. 347, by kind permission of McGraw-Hill Publishing Company.)

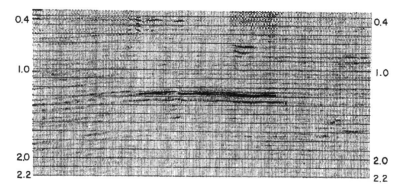

Figure 2.23 Record section designed to bring out relative amplitudes. This record is the same as that in Figure 2.22 but has been processed to preserve relative amplitudes. The strong reflection at 1.25 s is a so-called 'bright spot'. (Record reproduced from Milton and Dobrin, *Introduction to Geophysical Prospecting*, p. 346, by kind permission of McGraw-Hill Publishing Company.)

occasions where gas is present but not detectable. For example, it is not realistic to expect to detect gasified layers much thinner than 1–2 m, while peats and low grade coals can give a seismic record that appears similar to gas.

Figure 2.24 is a seismic record section where gas and liquids such as water and oil are in contact. Again the record has been processed to preserve relative amplitudes. Low energy reflections are shown alongside high amplitude events but the true relative amplitudes are distorted.[26] Figure 2.25 shows a well that actually blew out. Relative amplitude processing was employed to

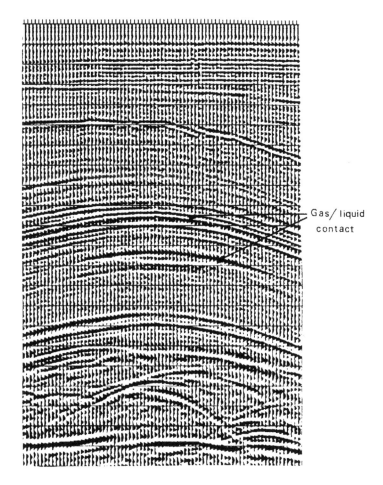

Figure 2.24 Seismic section showing gas/liquid contact in reservoirs. (Reproduced from McQuillan, Bacon and Barclay, *An Introduction to Seismic Interpretation*, p. 119 by kind permission of Kluwer Academic Publishers.)

show the amplitude anomaly but unfortunately this was not done before the rig blew out.[27]

Further progress can sometimes be made using other techniques. AVO (amplitude versus offset) analysis can be useful in certain cases, since the amplitude of a gas sand varies with offset in a way which is typically very different from other high amplitude reflectors. Forward modelling is another technique, where seismic response from an assumed model is compared with the actual response seen on the seismic record. A rule-of-thumb chart (Table 2.7) gives a qualitative guide to gas hazard assessment.[28]

It should, moreover, be emphasised that these characteristics apply to gasified sands in geologically young sand–shale sequences. Where gas is

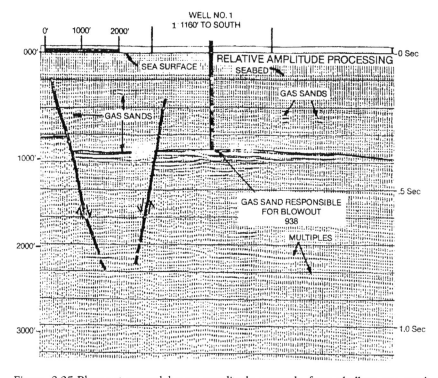

Figure 2.25 Blow-out caused by an amplitude anomaly from shallow gas–sand. (Reproduced from *Applied High Resolution Geophysical Methods* by Peter K. Trabant, p. 201 by kind permission of Kluwer Academic Publishers.)

Table 2.7 A qualitative guide to gas hazard assessment

Risk	Model probability	Typical characteristics
High	Gas probable	High amplitudes with three or four well defined features (positive AVO, closure, phase reversals, etc.)
Moderate	Gas likely	High amplitudes with two other subsidiary gas-like features
Low	Gas possible	Moderate amplitudes with one or two other features or very high amplitude alone
Slight	Gas unlikely	Usually one or more features but unremarkable amplitudes
None	Gas improbable	No evidence

Box 2.5
The philosopher's stone of geophysics

In 1972 the first gas accumulations were detected by seismic means. It is only a matter of time, thought many oil industry analysts, before seismic techniques will detect oil directly. Seismic surveys will give us all unlimited oil. This has never happened. Apart from anything else, bright spots have never been satisfactorily demonstrated on land. Some people think that the greater homogeneity of seismic transmission from water to the sub-sea-floor strata accounts for this, while others speculate that the fact that sound sources in water generate P-waves only is the answer. Either way the difference in transmission velocities between gas saturated sands and oil/water saturated sands is enough to identify bright spots, but the difference in seismic velocities between oil-saturated sands and water-saturated sands is less than 1 per cent and this is insufficient to differentiate between oil and water.

This author has two suggestions which combined might provide the basis for a solution. For the last 20 years the rapid development in the size and speed of computers has improved the accuracy of velocity fields, but even the largest fastest computers still give a velocity field that is not particularly accurate. A super-computer that worked for years on one velocity field would still not have defined it with total accuracy. To overcome this an alternative approach might suggest looking at the other end of the seismic package, namely the sound source and certain types of receiver. Let us suppose that a variable frequency sound source has a far field signature that can be varied on a line-by-line basis or perhaps a shotpoint-by-shotpoint basis. Also suppose that a 1200 m streamer has a second tailbuoy that could be towed 1200 m behind the usual tailbuoy. The second tailbuoy would have a sonobuoy attached with a hydrophone perhaps 20 m below the sea surface. A number of velocity control lines would be acquired, either the same line acquired several times with slightly different sound sources at each attempt, or a line with a different source signature at successive shotpoints. Each sound source would have a data set for a particular line or a number of velocity data sets for each shotpoint. If twelve slightly different source signatures were used with a 144-trace streamer, each shotpoint would have twelve data sets, one for each source signature. The sonobuoy would provide seismic refraction data which would be used for velocity control on the different source signatures. The average of the velocity fields for the different source signatures would be a highly accurate velocity field for a particular survey area. Could this differentiate between water- and oil-saturated sands?

78 *High resolution site surveys*

trapped within clays, the most obvious sign is usually widespread acoustic blanking, particularly noticeable on single trace analogue records. In any event, discharge rates in clays are likely to be very low, such that a gasified clay is much less of a hazard than a gasified sand. However, soil strengths of gasified clays are liable to severe reduction.

Digital interpretation is normally performed on a workstation. This has the advantage of speed in the construction of structural maps, plus the ability to examine attributes in colour. The most useful of these attributes are normally absolute amplitude, amplitude envelope and instantaneous phase. Colour displays are particularly useful.

2.15 Contractual work standards

The generally acceptable minimum standards for conducting digital site surveys are as follows.[29]

- The navigation computer should govern the synchronisation of all systems including shot intervals and data logging.
- The vessel speed over the ground should not exceed 6 knots, with 3.5–5.0 knots preferable.
- No more than 24 h of charged time should be allowed for deploying and checking out equipment. As much equipment checking should be done in port as possible and during the transit to the survey area. Equipment downtime should commence after 24 h of cumulative non-operation.
- Streamer polarity should be checked before the start of survey and after any change is made to the components or wiring of the array.
- Streamer depth indicators should be calibrated prior to carrying out the survey and at any time thereafter if it is considered necessary.
- Routine instrument checks should be carried out daily and faults should be corrected before continuing the survey programme.
- When more than one source is used, a means should be provided to check the alignment of individual pulses, and adjustments should be made as necessary.
- Only new magnetic tapes should be used and data from more than one line should not be recorded on the same tape.
- Work should not commence on any survey line if:
  - More than one trace is dead due to noise, polarity, leakage or insensitivity;
  - A streamer depth detector is bad or the allowable depth range is exceeded;
  - Streamer feather angles exceed 10° either side of the nominal line;
  - The energy source is low due to gun, generator or compressor failure;
  - At least two tape transports are not operable;

- Any monitor is inoperative;
- The positioning system is inoperable or inaccurate.
- Work should cease on a survey line if:
  - More than one streamer trace is bad, as described above;
  - A streamer depth indicator is bad, as described above;
  - Any equipment failure occurs, as described above;
  - There are more than eight consecutive misfires;
  - Any element in the positioning determination fails;
  - Positioning control exceeds ±5 m;
  - Vessel off-track deviation exceeds 10 per cent of the line separation.

References

1. Trabant, P.K. (1984) *Applied High Resolution Geophysical Methods*, International Human Resources Development Corporation, Boston, p. 138.
2. Texas Instruments, DFS IV Technical description.
3. Hatton, L., Worthington, M.H. and Makin, J. (1986) *Seismic Data Processing*, Blackwell Science, Oxford, pp. 47–48.
4. Texas Instruments, DFS V Technical Manual.
5. Sercel SN 358 Technical Manual, supplied by Fugro-Geoteam, Aberdeen.
6. TTS2 Technical Manuals supplied by TTS Systems.
7. OYO DAS-1A Manuals, supplied by Exploration Electronics.
8. Geometrics Strativisor NX systems description, provided by Geometrics.
9. Texas Instruments systems description and instrument tests. By the early 1980s most contractors had developed their own instrument tests. Several of these documents have contributed to this description.
10. TTS Manuals, supplied by TTS Systems.
11. Jones, E.J.W. (1999) *Marine Geophysics*, John Wiley and Sons, Chichester, p. 102.
12. Arthur, J. (1979) 'The application of high resolution multi-channel seismic techniques to offshore site investigation studies', *Offshore Site Investigation, Proceedings of a Conference Held in London*, March 1979, p. 77.
13. EG and G Sparker Technical Manual in the possession of the author.
14. Milton B. Dobrin (1976) *Introduction to Geophysical Prospecting*, McGraw-Hill Publishing Company, New York, p. 124. (Figures 2.14 and 2.15 reproduced by kind permission of the McGraw-Hill Company and Bolt Technology Corporation.)
15. *G. I. Gun. The Airgun that Controls Its Own Bubble*, technical manual, supplied by Exploration Electronics.
16. Parkinson, R. 'Quality control on seismic air-gun arrays', paper presented to a private BP seminar, circa 1982.
17. Jones, E.J.W., op. cit., p. 102.
18. Ibid., p. 104.
19. Litton LRS-100 technical data sheet.
20. *UKOOA Guidelines for the Conduct of Mobile Drilling Rig Site Surveys*, Volume 2, p. 37.
21. Digicourse Digibird 5010. Digicourse data sheet.
22. *Outline Procedures for Using Calibrated Hydrophones*, manual in the possession of the author, origin unknown, 15 pp.

23. Digicourse Digibird 5010 Digibird. Digicourse data sheet.
24. *UKOOA Guidelines for the Conduct of Mobile Drilling Rig Site Surveys*, Volume 2.
25. Milton B. Dobrin (1976) *Introduction to Exploration Geophysics*, McGraw-Hill, pp. 346–347.
26. McQuillan, R., Bacon, M. and Barclay, W. (1979) *An Introduction to Seismic Interpretation*, Graham and Trotman, London, p. 119.
27. Trabant, P. (1984) *Applied High Resolution Geophysical Methods*, International Human Resources Development Corporation, Boston, p. 201.
28. Gardline Surveys have used this chart for some years. Chart reproduced with Gardline's permission.
29. A fair summary of numerous oil company contracts in the possession of the author.

3 Analogue site survey systems

3.1 Categories of analogue systems

There are five main categories of survey system that form the basis of traditional shallow profiling site survey tools. These are echosounders, sidescan sonars, pinger/profilers, boomers, and multi-electrode sparkers. In addition there are a number of systems that combine two of these instruments in a single unit. Sidescan sonar and pinger combined units are one example of this. Each of these bottom and sub-bottom profilers is suited to different conditions and it is essential to know what survey information is required so that the optimum combination of equipment can be used. The different systems also have different weather capabilities and hence the time of year and area of operation will influence the choice of system. The type of shallow geology expected is also an important consideration which influences the optimum combination of systems. Technical developments in the 1990s included echosounder and sonar-based swath bathymetry. At the Millennium there was intense interest in autonomous underwater vehicles (AUVs) for deep water work.

3.2 Echosounders

The sinking of the *Titanic* in 1912 sparked the first sustained interest in a means of echo ranging to detect the presence of icebergs. Professor Fessenden produced the first working system and in 1914 interfering echoes from the sea floor as well as icebergs established the principle of depth determination by electronic means.

For site survey work a dual frequency hydrographic echosounder is considered essential. There are a variety of hydrographic echosounders in use, such as the Atlas-Deso 20 and the Odom Echotrac Fathometer. Figure 3.1 shows a schematic of the Atlas-Deso 20.[1] The echosounder should be correctable for vessel draft and velocity of sound in water. The echosounder transducer, mounted in the hull of the vessel, or on a vertical pole lashed to the side of the vessel, is energised to emit a pulse of high frequency sound, directed downwards in the form of a thin cone. This pulse travels through the water from the transducer and bounces off all surfaces which are more or less

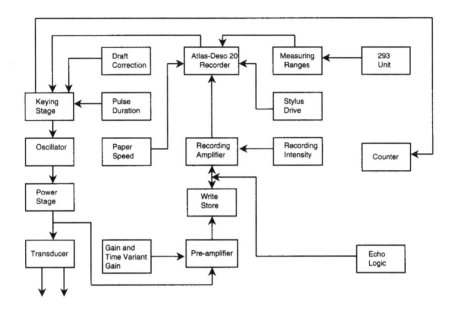

Figure 3.1 Atlas-Deso 20 schematic.

normal to the direction of propagation. The strength of the reflected signal depends on the reflectivity of the interface. In general the strongest and fastest consistent return is the return from the seabed vertically below the transducer. The return is charted as a particular time interval and the chart is calibrated such that the water depth may be read directly. Pulses are sent out at fixed intervals, between one and twenty pulses per second. Figure 3.2 shows the transmission and echo pulses represented diagramatically.

Navigational fixes are made at set intervals, typically every 100 m. The depth of the transducer must be known and the velocity of sound in water should be established. On most site surveys the echosounder data is digitised and recorded on the navigation data tape. Digitised depth data can be easily corrected for tidal variations. The echosounder is usually run during all phases of survey operation. If lines are shot twice, extra echosounder data coverage is obtained and this is an advantage since survey tracks cannot be exactly repeated twice.

A typical echosounder system consists of a transceiver unit, a transducer and a plotter. Modern systems tend to combine the transceiver and plotter along with all operating controls.

Quality control on echosounders

Quality control on echosounders involves checking the system parameters used for a particular survey. The power output and frequency range should be

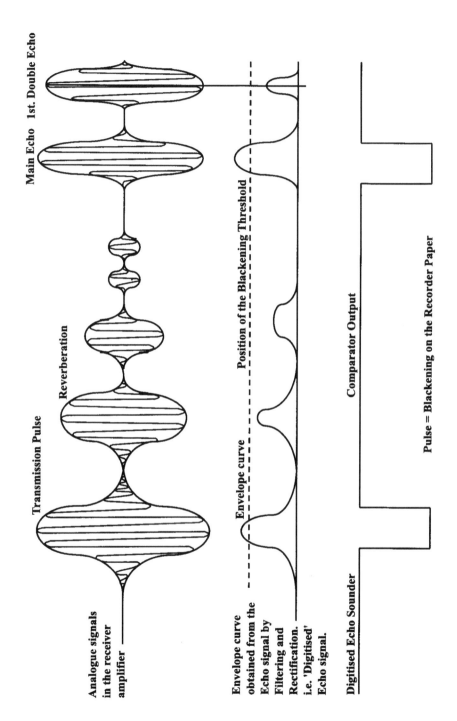

Figure 3.2 Atlas-Deso 20 input/output waveforms.

Box 3.1
Side-mounted echosounders

Many hastily mobilised site survey operations use vessels that do not have a hydrographic echosounder fitted as standard. The usual method of rectifying this is to secure the echosounder transducer to a bracket welded to the side of the ship. Time and time again this has led to disaster. On one survey, off the West coast of Africa, three successive transducers were lost due to floating logs striking the transducer mounting point on the side of the ship. On another the bracket itself was broken by the act of steaming to the survey area. Side-mounted echosounders are a disaster – and do not let anyone tell you differently.

established. Typical output frequencies are 33 kHz and 210 kHz. Generally speaking the 210 kHz transducer has a narrow beam width at 8.6° and the 33 kHz transducer has a broader elliptical beam with principal beam angles of 31.8° and 17.0°. The power output for these transducers is 170 W at 33 kHz and 150 W at 210 kHz. The resolution of the system should be in the region of 0.12 per cent, corresponding to ±12 cm in 100 m of water. Contractors' figures for beam widths should be treated with some caution as the criteria for beam width measurement are rarely provided.

Hydrographic echosounders should be adjustable for transducer depth and water velocity variations. Two methods exist for establishing the velocity of sound in water.

The bar check is the simplest method of establishing the velocity of sound in shallow water. A bar check also determines the exact position of the echosounder transducer below the vessel waterline and of course the vessel draft. A reflecting object is suspended from the vessel by a graduated chain so that it is illuminated by the transducer beam. A note of the actual depth is made and compared with the echosounder reading. The procedure is repeated for several depth measurements, typically at 5 m intervals. The average velocity error is calculated for the depth range required and this velocity adjustment is applied to the echosounder.

The second method uses a temperature/salinity meter on site at a succession of depths from the sea surface to the sea floor. This is always done on contemporary site surveys. The velocity conversion is performed using a number of slightly different formulae which have evolved over a period of years. A typical temperature/salinity meter is a Safre–Crouzet C10 velocimeter probe using Medwin's formula for velocity conversion. An alternative might be a Valeport 600 temperature/salinity meter using Wilson's modified formula to determine velocity. There are other meters and formulae. These formulae are empirical and have no definitive scientific basis. The Ocean Seven system measures not merely temperature and salinity but oxygen, pH, redox or NH_4^+ plus TILT.

Many echosounders are fitted with swell filters or at least with a swell filter option. A swell filter averages depth readings for a number of pulses to reduce the effects of swell in poor weather. The threshold should be considered carefully before applying the filter as some features may be smoothed out of existence. The swell filter is best left switched out until the weather gets so bad that depths cannot be easily averaged by eye from the paper readout. Typical echosounder parameters might be as follows:

- Transmit frequencies: 33 kHz and 210 kHz
- Transmit power: 170 W (33 kHz) and 150 W (210 kHz)
- Measurement accuracy: 0.12 per cent (9.5 cm) for 33 kHz, 0.12 per cent (1.5 cm) for 210 kHz
- Heave compensation: Model TSS 320
- Calibration: Safre–Crouzet C10 velocimeter using Medwin's formula

Having established the system parameters, noise evaluation forms the basis of most quality control. Figure 3.3 shows the effect of leakage in the transducer. The bottom profile is still evident but all detail is lost and the records are not acceptable. Figure 3.4 shows the record obtained from the same system after it had been properly set up.

In conclusion, the echosounder is a reliable method of obtaining bathymetry for high resolution site surveys and pipeline route surveys. If carefully calibrated and backed up with good tidal corrections it can provide accurate detailed bathymetry for platforms, especially concrete platforms. In suitable conditions of shallow water and soft formations the echosounder can penetrate the subsurface to a few feet to give a certain amount of detailed shallow geological information.

The correct use of the wide beam transducer allows a degree of search ability from the echosounder. Wellheads can be found with a sidescan sonar and pin-pointed with the echosounder. The weather dependability of the echosounder is a function of the information required since the record will show faithfully the movement of the vessel and an average profile can always be drawn. The point at which the sea becomes too rough for bathymetry work is therefore a qualitative decision in the hands of the onboard survey personnel.

Echosounder data should be processed using a heave compensator to remove vessel heave caused by wave action or swell. Data should be reduced to least astronomical tide (LAT) or mean sea level (MSL).

3.3 Sidescan sonars

A sidescan sonar is designed to map the surface tomography of the sea floor. Sidescan sonars were developed in the 1950s from experiments using echosounder transducers tilted away from the vertical. These early experiments

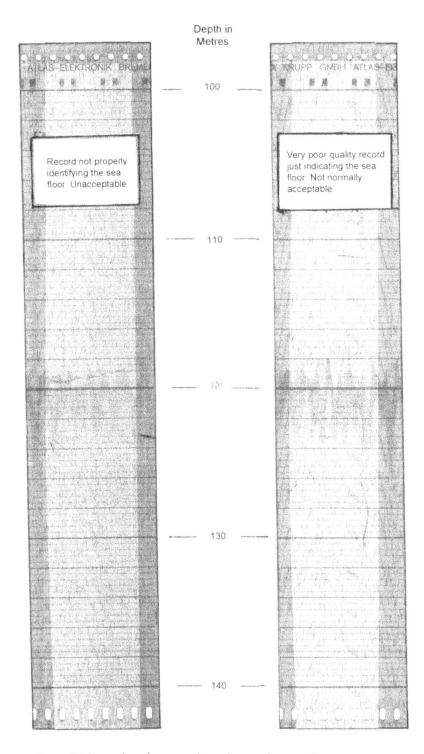

Figure 3.3 Examples of poor quality echosounder record.

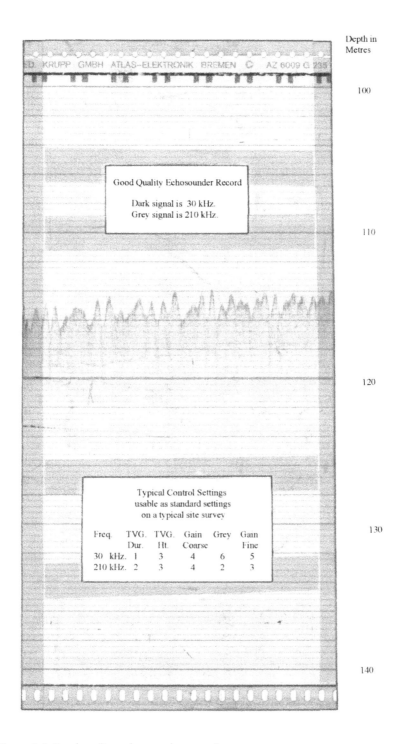

Figure 3.4 Good quality echosounder record.

were conducted in Britain as an offshoot of the World War II anti-submarine research programmes. The results of these experiments showed the potential of this method as a tool for the study of natural seabed features not directly beneath the survey vessel. A typical sidescan sonar consists of a towfish fitted with the two transducers, a receiving amplifier and signal processor, a dual trace printer and a tape recording system. Figure 3.5 shows a schematic of a typical sidescan sonar system.

The resolution of sidescan sonars is primarily controlled by the pulse length, pulse repetition rate and the beam width of the transducer. Of equal importance are the fish height above the seabed and the range in use. Sonar fish movements such as pitch and yaw further degrade the resolution of the system. Resolution can be divided into transverse (perpendicular to track) and axial (along track) errors. Transverse resolution depends on pulse length. A pulse length of 0.2 ms should give a range resolution of 15 cm, while a pulse length of 0.01 ms will give a range resolution of 0.75 cm. Due to slant range distortion these figures will only be correct at extreme sonar ranges where the angle of incidence is close to the horizontal. Axial resolution is controlled by the beam width and the pulse repetition rate, combined with the vessel ground speed. Beam width along track is determined by the sonar transducer, and as with echosounder beam widths, manufacturers' figures should be treated with caution. A beam width at 2° should cover a slant range of 349 cm at 100 m, while a 5° beam width should cover a slant range of 746 cm at 200 m. Sonar slant range correction and speed correction principles are shown in Figures 3.6 and 3.7.

The principal feature of the towfish is the ability to emit a focused beam of sound having a narrow horizontal beam angle, usually of 2° extent, and a vertical beam angle of 20°. The acoustic pulse is of short duration, typically 0.1 ms, initiated from the recorder and emitted by the fish. The acoustic frequency used is usually 105 kHz. The towfish uses dual channels so that it can scan to either side of the ship, illuminating a swath perhaps 200 m to either side of the vessel. The detected signals are presented on a facsimile record. The record presentation is such that the time scan can be calibrated in terms of distance across the seabed. The first echo in any scan is the bottom echo. Any subsequent echoes are reflected from features ranging across the seabed to the outer limit of the scan. Most sidescan systems are designed for operation in continental shelf areas for work in depths up to 250 m. The altitude of the towfish is read directly from the sidescan record as the first reflection from the closest point on the seabed, directly below the towfish. Depth control is maintained by varying the vessel speed or altering the tow cable length. The height of the towfish above the seabed should be approximately 10–15 per cent of the selected range. Modern sidescan sonar towfish include the EG and G Model 272 towfish, while the EG and G Model 990 towfish can operate to depths of 6000 m. This unit uses a lower transmit frequency at 59 kHz.

A modern sidescan image correcting system will give a plan view of the sea floor, corrected for vessel speed, towfish height and signal amplitude. Modern

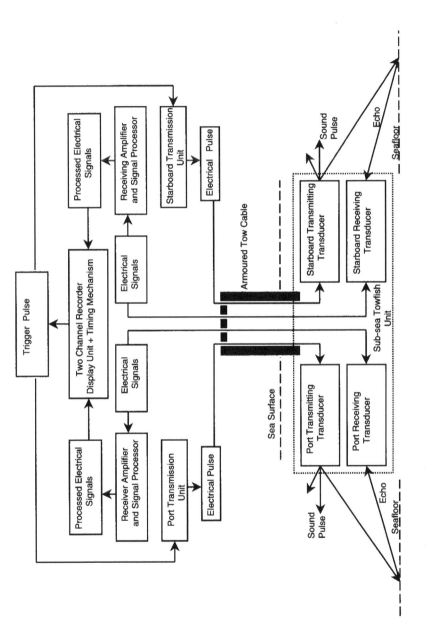

Figure 3.5 Sidescan sonar schematic.

90   *High resolution site surveys*

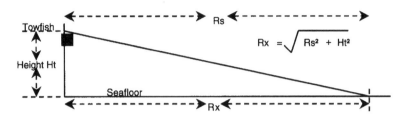

Figure 3.6 Slant range correction principles.

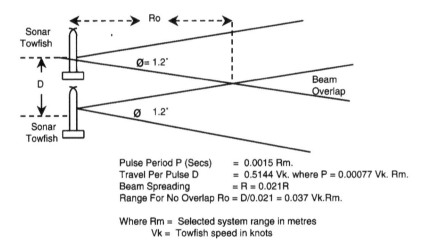

Figure 3.7 Sonar speed correction principles. Since beam spreading occurs both vertically and horizontally there will be no loss of coverage when R_o equals the fish height. For example, when the towfish height is 20 per cent of the range, there will be no loss of coverage for speeds of less than 5.4 knots.

recorders will be microprocessor-based, dual channel hard copy display units with the data digitised for further processing either onboard the survey vessel or by geophysical processors at a later date.

Typical sidescan sonars include the now elderly EG and G Mk 1B system and a very similar Klein system, Model 531T recorder with Model 101 AF towfish. The much more modern EG and G 260-TH image correcting system is a better option for contemporary surveys. The Geoacoustics 159 or 160 digital sidescan sonars are similar contemporary systems. Another new development is the Geoacoustics dual frequency sidescan sonar. In this system two transmit frequencies at 100 kHz and 410 kHz give high or very high resolution.[2] Table 3.1 attempts to compare the sidescan sonars available today.[3]

Table 3.1 Comparison table of sidescan sonars available today

Company	System	Towfish depth (m)	Frequency (kHz)	Pulse length (ms)	Power (dB)
C-Max	CM800	500	325	53–160	217
Datasonics	SIS-3000	6500	55–65	1–40	223
Datasonics	SIS-1000	1000	90–110	1–24	226
Datasonics	SIS-1500	1000	190–210	1–24	226
Geo-Acoustics	159D	2000	114	0–166	220
Klein	SSS	1000	100	0.025–0.5	228
Marimatech	E-Sea Scan 800	500 or 1500	325	50–160	220
Micrel	MICM 800	1500	325	1.07	217
Odom	Echotrac Mk.11	30	200	Variable	225
Odom	Multibeam	10	200	Variable	225
Submetrix	ISIS-100	300	234	0.08–2	220
Ultra	Widescan 3	1500	100	0.1–0.5	220
Ultra	Deepscan 60	6000	60	Variable	231

The power figures are given in dB relative to µPa at 1 m.

Typical sidescan sonar site survey operating parameters might be as follows:

- Beam depression: 10°
- Vertical beamwidth: 20°
- Horizontal beamwidth: 1°–2°
- Range setting: 200 m/channel
- Towfish operating frequency: 105 kHz ± 10 kHz
- Pulse duration: 0.1 ms
- Power output: 128 dB ref. 1 µbar at 1 m
- Towfish depth: 90 per cent of operating water depth
- Pulse rate: Six pulses per second

There is increasing interest in very deep-tow sonar systems. In such systems the towfish may be at the end of as much as 9000 m of cable. The uncertainty of the towfish position may be such that it will be dynamically positioned using an acoustic transponder. An acoustic pinger will be mounted on the towfish and a chase boat will be fitted with an ultra short baseline (USBL) acoustic receiver. A second vessel is necessary because the towfish may be as much as 5–8 km behind the survey vessel. Typically, the transponder on a deep-water towfish will have a power output of 208 dB and a narrow beamwidth of perhaps 15°. Systems such as these will have a pulse trigger derived from the differential mode GPS positioning system. The transponder on the towfish is then triggered directly by the DGPS and the acoustic positioning pulses are received by the chase boat. The position of the towfish is then computed and displayed. The positioning system data from the chase boat is telemetered to the survey vessel and the towfish positioned to an accuracy of perhaps ±3 m. The acoustic positioning systems necessary to do this are reviewed in Chapter 5.12 of this book. There are other problems that manifest themselves when very long sonar tow-cables are used. Comment has already been passed in Section 1.7 that at the end of a survey line the turn-around to the next line is greatly extended.

Quality control on sidescan sonars

Quality control on sidescan sonars is relatively straightforward. Provided the towfish electronics are operating satisfactorily then the height of the towfish above the seabed needs to be checked. If the fish is too high then sea floor detail may be lost. Contractors have a tendency to run the fish too high to minimise the risk of losing the towfish when it strikes the bottom due to a decrease in the ship's speed or a sudden rise in the seabed. The following general requirements represent a reasonable approach to quality control on scanned sonar systems.

Box 3.2
1001 ways to lose a sidescan sonar towfish

The most common accident on site survey operations is the loss of the sidescan sonar towfish when it strikes the seabed. It is actually very difficult to avoid this. The towfish will dive to the seabed under almost any pretext. To give three examples, the survey ship slows when heading up into wind and sea and the officer of the watch does not increase vessel speed quickly enough. Second, the seabed may rise sharply and the sidescan operator does not reel in the towing cable quickly enough. Third, the towfish may hit a mid-water obstruction such as fishing nets or become entangled in other debris and detritus in the water. The usual outcome is that the safety link snaps and the towfish reverses its orientation in the water, having performed a backflip, which hopefully disengages the towfish from the obstruction. It can then be winched in and repaired as necessary. If the towing cable is cut, usually where the cable splice meets the towfish, the towfish will be lost completely and most survey operations carry two or three towfish units on the assumption that one will have a good chance of being lost. When the towfish is in the water a sidescan operator should always be supervising its position in the water and depth above the seabed. This is an instrument that should not be left unattended.

- The ideal height of the towfish above the seabed should be 10 per cent of the operating range. Any prominent features identified by the sonar should be revisited and resurveyed at the shortest useable range.
- The layback of the towfish should be accurately known. On deep-tow sidescan sonars this may entail the use of an acoustic positioning system.

- Instrument adjustments should always be made before the start of a survey line. Amplifier gains should not, as a general rule, be varied during the acquisition of a survey line.
- Crosstalk between the port and starboard channels should be reduced to an absolute minimum.
- Interference from onboard electrical equipment should be checked and eliminated.
- According to the type of sonar used, the sonar transducers should be adjusted for optimum beam and depression angles and these values should be logged.

The recorder is always a dual channel unit which should have filtering, overall gain and time variant gain (TVG) control on each channel. A minimum amount of filtering only is necessary due to the fact that the source is of a fixed frequency. The TVG controls are very important and the survey personnel must ensure that they are correctly adjusted in order to give an even gain intensity across the entire record. Most contemporary systems incorporate digital tape recorders. The EG and G Model 360A dual cartridge tape recorder is a good example. This unit uses two tape drives with automatic switchover to provide continuous recording of sonar data. The recorder duplicates the hard copy visual record, corrects graphic presentations due to navigation errors, layback and towfish position and improves image resolution through the use of image processing systems. This increases the effective dynamic range and improves resolution. The sonar data can be processed onboard and amplitude manipulation can be carried out if desired. Scale distortion can also be removed.

Figure 3.8 shows a section of good quality sidescan sonar record, with the towfish positioned rather too far above the seabed. The light areas show smooth sands, while the dark patches are clays with gravel. The water surface reflection shows only faintly while the seabed reflection is very strong indeed. The range is indicated in metres and shows a system set up for a range of 125 m on each channel.

Figure 3.9 is of particular interest and shows a well that has blown out. The Sonargraph survey was conducted some months after the blow-out and shows a jagged hole in the sea floor, with gas still seeping into the water column. In the case of massive water aeration by gas the survey vessel buoyancy becomes a serious consideration when data acquisition is considered.

Figure 3.10 shows some of the interference patterns which must be diagnosed and corrected in the field. Interference from other survey systems such as multi-electrode sparker (MES), boomer or profiler is usually due to the method of deployment of the systems such that the sidescan tow cable is run too close to the electrical cable associated with capacitor discharge systems in use. Winch motor vibrations on the slip rings is a common source of interference. Towfish instability is a sign that one or both the stabilising fins have been lost, or that the fish is dragging on something such as an old cable or fishing net.

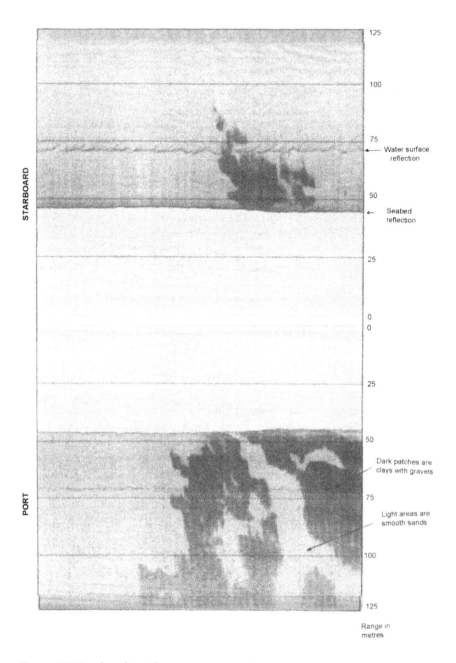

Figure 3.8 Good quality sidescan sonar record.

Analogue site survey systems 95

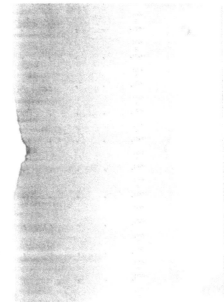

Figure 3.9 Wellhead blow-out seen on a sidescan sonar record.

The sidescan sonar is an invaluable search tool used to locate wrecks and other obstructions on the sea floor. It is essential in pipeline surveys, rig site surveys and platform location surveys. With care it can show boulders down to a diameter of less than 0.5 m. It can also be used to detect wellheads, drill pipes and casings, rig anchors and other solid obstructions.

3.4 Sonar swath bathymetry

The sonar swath bathymetry technique allows the collection of large amounts of bathymetric and sonar data from a relatively small number of passes across a survey area. Such techniques are widely used for pipeline inspection surveys and for acquiring low cost bathymetric data in areas where chartage is poor. There are two principal methods for acquiring swath bathymetric data. A multibeam echosounder is one technique, while the second uses what is effectively a modified sidescan sonar utilising phase measurement techniques to determine seabed depths across the survey vessel track.[4]

In the early 1980s it was realised that multiple beam focused sonars could be built that removed the slant range ambiguity due to sloping sea floor features. The result was systems which produced a complete topographic map where the elevation and position of conjugate features was obtained from parallax measurements rather as in photogrammetry. This did not alter the fact that along track resolution is inversely proportional to range, due to

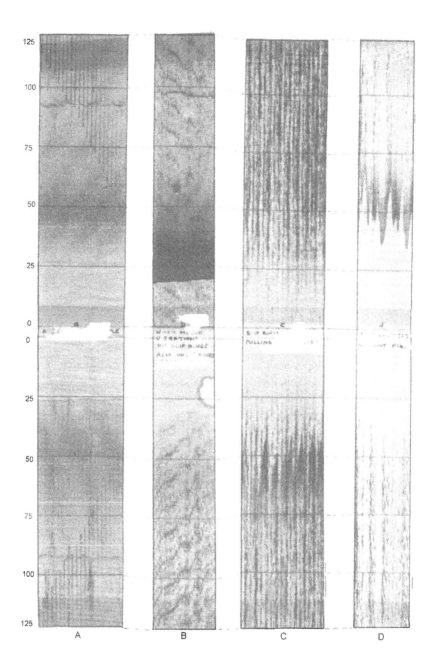

Figure 3.10 Sidescan sonar interference patterns. (A) Multi-electrode sparker interference. (B) Winch motor vibrations on slip-rings, also unbalanced gain. (C) Slip-ring interference while pulling in the towfish. (D) Towfish unstable without fins.

spreading of the beam. It should also be understood that range ambiguities limit the pulse repetition rate. Area coverage rate and range resolution are always in conflict and an improvement in one characteristic is brought at the expense of the other.

Synthetic aperture radars have been used by the military for decades. The idea is always to stack (add coherently) the transmissions from a small aperture (wide beam angle) moving source. The reflections received from targets will be doppler shifted by an amount dependent on the target azimuth and the velocity of the source/receiver. As the sonar source moves past the target the doppler shift will change from positive to negative and the shifts can be measured by phase coherent processing of the received signals. Target range and azimuth can then be established.[5] It is only in the last few years that these radar principles have been applied to sonar. Difficulties with range ambiguities caused by the transit time of signals from distant targets delayed development of these systems. The survey vessel moves significantly between transmissions due to the low speed of sound in water. Ship stability, position errors, currents and wave action, etc. all affect the accuracy of results. Nevertheless, swath bathymetry is now a practical proposition. Figures 3.11 and 3.12 show schematics of a swath bathymetry system. A functional block diagram and a software block diagram indicate that the system's software is as important as the actual engineering of the system.

Echosounder-based swath bathymetry

Multibeam echosounders consist of several transducers mounted in a single unit. The transducer unit can be survey vessel hull mounted, deployed from a remote operated vehicle (ROV) or from an autonomous underwater vehicle (AUV). There are obvious advantages in using a hull-mounted unit because the transducer position can always be known exactly and the choice of a vessel with a hull-mounted transducer unit can be the critical determining element in the success or failure of the survey. ROV and AUV-mounted transducers can have the advantage of very high density soundings over small areas, such as pipeline freespans.

A typical modern multibeam echosounder is the Simrad EM 300 which can operate from a water depth of 10 m to water depths of 5000 m with swath widths of 6000 m. A nominal sonar frequency of 30 kHz with an angular coverage sector of 150° is used. There are 137 simultaneous beams, $1 \times 1°$, $1 \times 2°$, $2 \times 2°$, or optionally $2 \times 4°$ beams and a range sampling interval of 15 cm. The angular coverage sector and beam pointing angles are variable with depth according to achievable coverage, to always maximise the number of useable beams. The beam spacing is normally equidistant, corresponding to 1.5 per cent of depth at 90° angular coverage, 2.5 per cent at 120°, 4 per cent at 140° and 5.5 per cent at 150°. The transmit fan is split into several individual sectors with independent active steering according to vessel roll, pitch and yaw to obtain a best fit to a line perpendicular to the survey line and

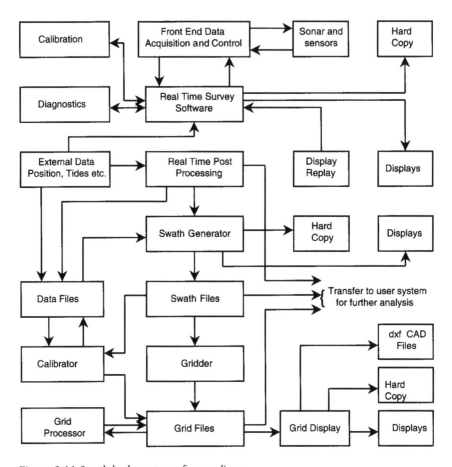

Figure 3.11 Swath bathymetry software diagram.

thus a uniform sampling of the bottom. The sectors are frequency coded (30–34 kHz) and are all transmitted sequentially at each ping. Vessel steering is fully taken into account when the position and depth of each sounding is calculated as is the sound speed profile effect on ray bending. Pulse length and range sampling rate are also variable with water depth for optimum resolution. The ping rate is limited by the round trip travel time up to 10 Hz. System accuracy is usually stated at 15 cm or 0.2 per cent of the depth RMS. The EM 300 transducers are linear arrays with separate units for transmit and receive.[6]

Another typical swath bathymetry system is the Elac Bottomchart MK11. In this system the coverage is 7.5 times the water depth, to a depth of 400 m (±7.5° beam direction with reference to the vertical) and 3.5 times the water depth to 1200 m depth (±60°).

Figure 3.13 shows a graph of water depth versus swath coverage.

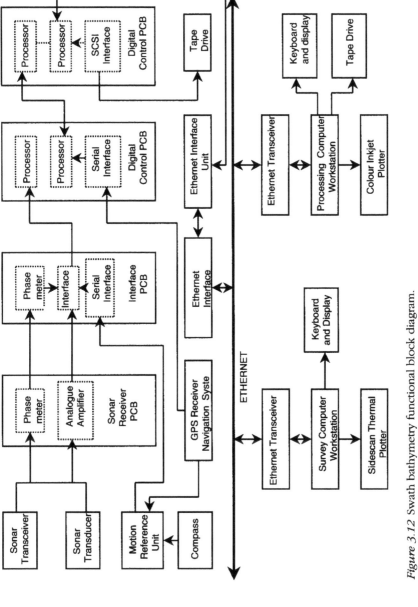

Figure 3.12 Swath bathymetry functional block diagram.

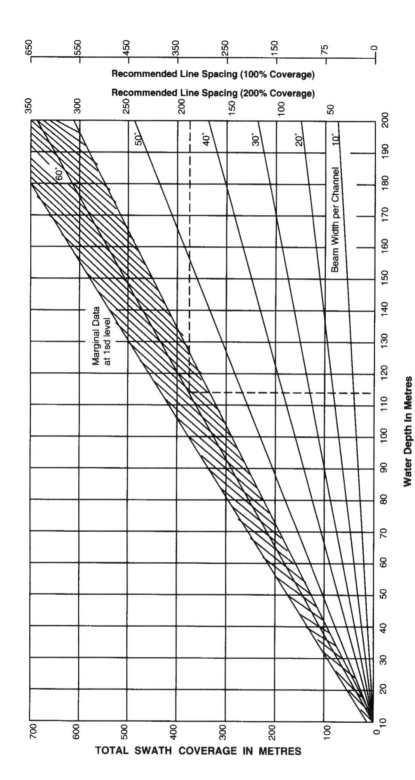

Figure 3.13 Graph of water depth versus swath coverage.

Sonar-based swath bathymetry

The phase measurement technique usually consists of a sidescan sonar-based system with pairs of transducers mounted each side of a towfish, one transducer in each pair being mounted horizontally above the other with a known separation distance. A short pulse of sound of a known fixed frequency is transmitted and this pulse follows a beam shape that is narrow in the in-line direction and broad in the cross-course direction. The seabed backscatter is recorded as a function of elapsed time as in a conventional sidescan sonar system. Where the swath system differs from a conventional sidescan system is that the phase characteristics of the seabed return, measured between each pair of transducers, is recorded instead of the amplitude. With a phase measurement system the entire seabed below the transducers is ensonified by each pulse and discrete sections of the returned signals can be analysed. This can result in many more soundings being measured across the ship's track in a single pulse than with a multibeam system. This process is known as interferometry due to in-phase signals producing interference patterns across the seabed.

A typical modern sonar swath system is the Submetrix ISIS 100 interferometric seabed inspection sonar system. The main system components are the sonar transducers, the electronics rack system, a motion reference unit and a computer workstation. The sonar transducers can be fixed to the vessel hull or pole mounted from the hull. By measuring the motion and location of the transducers the depth information is correctly located with respect to the Earth's surface. The seabed can be inspected while the survey is underway and post-processed data gives a digital terrain map (DTM). Two operating frequencies are available, 234 kHz for 0–100 m water depth operation and 117 kHz for 0–200 m water depth operation. The sounding repetition rates are variable, to give nominal ranges from 70 m to 300 m. This gives swath widths of twice these ranges and corresponds to ping repetition frequencies (PRF) of 10.7 and 2.5 times per second respectively. Roll, pitch, heave and heading are all compensated for and measured to an accuracy of 0.05° within a range of ±50°. ISIS data is recorded in the Submetrix Generic Data format which is based on a UNIX 'tar' format. The final result is sonar data amplitude sampled every 35 µs, giving 2800 pixels on a 150 m range sweep. This is as good as any conventional sidescan sonar system. Bathymetry data is recorded simultaneously.[7]

Another typical sonar swath system is the Geoacoustics Geoswath system. It comes in two versions which operate at 125 kHz and 250 kHz. It can work in water depths up to 200 m with a maximum swath width of 600 m. The beam width of the transducers is 1.7° of azimuth. The transmit pulse length is 100 µs to 1 ms. The number of swaths per second depends on the swath width.[8]

Swath bathymetry discussion

In sum, the echosounder-based systems have the advantage of hull-mounted transducers whose position is always accurately defined, while the sonar phase-measurement technique gives 100 per cent ensonification from a less accurately defined position derived from a towfish-based system.

In terms of data acquisition, the modified echosounder technique and the phase-measurement sonar interferometry method both produce a three-dimensional fan shape of acoustic energy that is transmitted from and received by the transducers. This fan shape is subjected to pitch, roll, yaw and vertical oscillation, all of which must be compensated for in processing and the time flagging between all the sensor data must be established. In addition the timing of depth data with positional data is critical. The speed of sound in water must also be known. Quality control is based around the measurement of these various parameters.

Gyro calibrations

The measurement of yaw requires an accurate gyro calibration. This usually involves mounting two prism targets at the bow and stern of the survey vessel on an exact centre-line. Gyro compass readings onboard the vessel are then synchronised and compared to range and bearing measurements taken to the bow and stern targets from a co-ordinated reference point on shore, usually the quayside. The vessel is then turned through 180° and the process repeated after the gyro has settled. Such gyro calibrations are normal for 3D surveys but all too rare for site surveys. On site, a gyro calibration can be undertaken by using a wellhead or other prominent feature. The chosen feature should be midway between two parallel survey lines, with the lines acquired in opposite directions. Each line should show the object and any spatial errors can be corrected with a calculated gyro offset.[9]

Pitch and roll offset calibrations

On-site gyro calibrations are only valid if pitch and roll offset calibrations have been performed. Time delay (pitch) calibrations should be performed over a steep slope. Ideally a single survey line should be acquired perpendicular to the slope. Two calibration runs should be made, one in each direction, and a difference correction factor calculated. A third attempt at the calibration line should confirm the calibration results.

Roll calibrations should ideally be conducted over a flat seabed. Two perpendicular survey lines should be acquired, with the centre beam from one survey line in exactly the same position as a single sweep of data from the line in the perpendicular direction. Comparison of the cross profiles allows the roll offset to be calculated. For onboard quality control a single line can be acquired in opposite directions, with the centre beams from the two lines as close as possible from the survey line. This result can then be compared with the first calibration technique and the first calibration confirmed.

Swath system calibration using an echosounder

On site the simplest test of a swath bathymetry system is to reference it against a conventional echosounder, preferably along a designated survey line. Depths should agree within the tolerances of the two systems. Roll calibrations are best performed over a flat seabed. Two perpendicular lines should be acquired with the centre beam from one survey line in the same position as a single data sweep from the perpendicular line. Comparison of the cross profiles determines the roll offset. Pitch and offset timing should be performed over a sloping seabed with a calibration line perpendicular to the slope acquired in two directions. A difference chart can be derived from the two passes and a correction factor applied. A third pass can then be conducted with corrections applied to confirm the calibration.

Further swath system calibrations

Further calibrations for gyro offset, alignment and timing errors can be performed over a known seabed feature such as a wellhead. Two parallel lines can be plotted with the feature at the midpoint. The two passes can then be compared and any spatial errors can be corrected by applying a gyro offset, provided that pitch and roll calculations have already been performed. Finally, calibrations for vessel draught change due to speed can be made at different speeds, for example over a speed range of 2–6 knots, and an assessment of data acquired while the survey vessel is manoeuvring. Manoeuvring data can be acquired from two slightly overlapping circles with the points of intersection joined by sailing a straight line. The three data sets can then be integrated and compared. An additional calibration for assessment of ray-bending correction and sound velocity profile is sometimes performed over an even slope with two perpendicular lines at 45°, forming a cross. Contours should be straight to the limits of the swath.

As a general rule there should be a swath overlap of 5–35 per cent with a reasonable number of cross lines. Typical operating parameters should be as follows:

- Vessel speed: 4–6 knots
- Operating depth: 5–1000 m
- Number of beams: 48–96, perhaps 120
- Sector: 60–150°
- Beamformer: Digital
- Frequency: >50 kHz
- Resolution: 0.05 m

Swath bathymetry data processing

Sonar swath bathymetry data is usually subjected to much more extensive processing than conventional sidescan sonar data.[10] A typical basic processing sequence might include the following:

1. Navigation post-processing
2. Raw data
3. Application of tidal data, water column data (temperature and salinity), depth corrections, etc.
4. Despiking
5. Filtering
6. Binning and gridding
7. Contouring
8. Plotting

Raw data can be assessed using colour coverage plots which clearly indicate the number of soundings used. An along track profile, based on the centre beam of the system, can be produced and compared with the echosounder data. The corrections can then be applied for the reduction of tidal data to LAT (least astronomical tide) and for vessel-induced errors such as gyro, squat, timing, heave and pitch/roll.

Data editing usually involves a first pass spike editor, with no valid data removed. Filtering usually consists of beam editing and the evaluation of data quality may ensure that several of the outer beams are edited. The filtering of binned data may also be necessary, based on statistical analysis. The removal of depth spikes usually requires a two-pass filter. The first pass filter is usually $3 \times SD$ and the second pass filter $2 \times SD$. The size of each bin in the final binning grid will depend on the survey in question. Equally the contouring is variable according to the survey. The final charts for a site survey are usually 1 : 10,000 with smaller scales for pipeline route surveys and debris clearance surveys.

3.5 Autonomous underwater vehicles

At present there is considerable interest in remote-controlled autonomous survey vehicles that are acoustically positioned and can carry out survey work independently of the towing requirement of the survey vessel. The very long line turns necessary with 2500–3500 m of towing cable deployed have already been noted in Chapters 1.7 and 3.6 of this book. Obviously, huge advantages accrue to the user of an effective AUV in deep water. Such a vehicle will usually give bathymetry (depth and position), using a swath echosounder. The sea floor will be mapped by conventional sidescan sonar and the sub-bottom sediments will be mapped to a depth of perhaps 30 m using a pinger transducer.

The multibeam swath sonars to be fitted to AUVs will probably operate in the 200–450 kHz range, with a swath width of 200–300 m. Up to 128 beams will be used for swath bathymetry and ensonification of the target area. If conventional sidescan sonars are used they are likely to operate in the 90–210 kHz range and perhaps in the 290–675 kHz range with power outputs in the 200–230 dB, reference 1 µPa at 1 m range. Horizontal beamwidths will be in the 0.5° to 1.2° range, with 50° to 60° of vertical beamwidth. Sub-bottom profilers will operate in the 400 Hz–24 kHz range, with penetration to as much as 100–200 m in silt-clays but perhaps only 10–20 m in coarse sands. It is likely that magnetometers and cable trackers will also be fitted, in addition to laser linescan cameras, still cameras and video cameras.

With towed systems the position of the towfish in very deep water will be ever more uncertain as the length of cable increases. It does not matter if the towfish is autonomous or towed from an umbilical, beyond a certain point, probably about 500 m of water depth, the towfish or other vehicle should be acoustically positioned. At the time of writing it is uncertain how much information will be required from the autonomous vehicle onboard the survey vessel and how much data will be recorded onboard the vehicle for downloading at the completion of each diving mission. It is probable that regular telemetered positioning data and quality control monitoring of the geophysical data would be sent from the AUV to the survey vessel. The main database would then be transferred at mission completion.

If AUVs are to be properly developed there are a number of technical problems to be overcome. They will almost certainly require the larger type of survey vessel as a deployment and recovery base. Stern handling with an A-frame or crane may be too risky. Amidships handling using a 'moonpool' may be necessary. A large proportion of the AUV's power will be devoted to getting down to the target depth and back up again. Drop weights may help to get the vehicle down to the sea floor without too much power loss. In bad weather an AUV might be at an advantage over a towed vehicle. The AUV would merely continue surveying and if necessary be 'sent to sleep' at the end of mission until the weather moderated sufficiently to allow recovery. The AUV should be fitted with collision avoidance sonar but inevitably it would at some stage tangle with mid-water floating driftnets, which are invisible to sonar. It should also be realised that through-water video data transfer does not exist at present and the technology to do this lies some decades in the future, if it is ever developed at all. It is probable that an effective laser linescan camera would be required for the AUV and the AUV should be capable of carrying a number of survey packages that would vary from survey to survey. The following is an attempt to summarise the AUVs that are currently in existence or projected.[11]

Martin 200

The Martin 200 is built by Maridan AS of Denmark, has a 1000 m design depth with an endurance of 200 km at 4 knots. This system carries a multibeam echosounder, a sidescan sonar, a sub-bottom profiler, a laser-scan camera and a video. The vehicle uses the MarCIP (Communication, Interface and Power) system, which is a combined underwater power supply and data communications system. This supplies power to all the instrumentation onboard the vehicle. The vehicle also uses MarCom which is a software system for data communications in time dependent applications.

The Martin vehicle is controlled from the survey vessel using acoustic communications. The range of the system is about 2 km and utilises the Orca Instrumentation MATS system, which adapts to the acoustic environment by shifting between low and high data rates. The low data rate is 300 bit/s using frequency shift keying or 2400 bit/s using phase shift keying. Downlink data to the vehicle includes commands that relate to the survey plan or replies to fault status messages. Uplink data includes vehicle position and status. The vehicle computer allows it to follow a pre-programmed survey route. The positioning system is based on the RDI Workhorse Doppler velocity log and an inertial platform fitted with ring laser gyros. Position fixing is by dead reckoning based on the Doppler velocity log and integrated navigation system data. Loss of positioning accuracy is 5–7 m/h.

The system is recovered by a drag net lowered into the water. The vehicle can then be steered or steer itself into the centre of the recovery net. Other conjectured recovery methods include a docking station that can be lowered to the seabed.

Hugin

The Hugin project is a collaboration between NDRE, Kongsberg-Simrad and Norwegian Underwater Intervention (NUI). The Hugin vehicle is built by Kongsberg-Simrad with funding from Statoil and various other companies. The vehicle is 4.8 m long and has a design depth of 600 m with a 3000 m version under development. It is fitted with a multibeam echosounder. Hugin is an operational system and conducted its first commercial survey in 1997.

Hugin operates and navigates according to a pre-programmed mission plan. Normally a mission plan is based on a pre-survey assessment of the area in question which is downloaded to the AUV before it is deployed. The vehicle establishes a set course and depth where it performs a scheduled sequence of mission steps. Once at a stable height above the seabed, multibeam data of constant swath width is acquired. Acoustic communications allow the operator onboard the survey vessel to modify operational and survey plans as required. System status messages are continually telemetered to the survey ship and in a crisis the AUV can be surfaced either from its own automatic command system or from an instruction from the survey vessel.[12]

Hugin is positioned by integrating the high precision acoustic positioning (HiPAP) system and a differential global positioning system (DGPS). The HiPAP system tracks the AUV with an angle measurement accuracy down to 0.2°. The HiPAP system uses a multi-element spherical transducer design that tracks the AUV with a narrow beam, using electronic beam-steering control. A total position accuracy for the multibeam echosounder position can be achieved to better than 0.45 per cent.

Hugin is deployed from a 30 ft container located at the stern of the survey vessel. The vehicle returns to the surface using its propulsion and rudder system. There is an emergency ascent sequence that involves the release of a ballast weight and the inflation of an internal bladder. On reaching the surface the vehicle drops its nose cone and releases a retrieval line which is then recovered by the vessel crew. The AUV is then hauled up a special ramp recovery system projecting into the water from the vessel container.

International Submarine Engineering – Theseus

International Submarine Engineering (ISE) and its wholly owned subsidiary ISE Research are a Canadian company specialising in the design of remotely operated vehicles, robotic systems for nuclear applications and lately, of AUVs for the oil industry. ISE have developed three AUVs, Dolphin, 'ARCS' and Theseus. Theseus has a direct defence capability, specifically it can lay long lengths of fibre optic cable under the Arctic pack ice. Theseus can carry a wide range of survey systems, can be remotely controlled or run to a pre-programmed survey plan. Theseus is 10.7 m long and 1.3 m in diameter. Theseus can navigate over significant distances, a maximum of 500 km being claimed, and can be sent commands using acoustic telemetry. Its speed is 4 knots. It can operate at depths up to 1000 m.

Bluefin Robotics/MIT

Odyssey is built by Bluefin Robotics/MIT. The fish is 2.15 m in length and can operate to 6000 m. It is mainly used for oceanographic work. It is fitted with water quality monitoring sensors, sidescan sonar and a current measuring device. The endurance is 12 h at 5 km/h.

Autosub

Autosub is one of two United Kingdom contenders in this market. The vehicle is 6.82 m in length and has been developed by the Southampton Oceanography centre, mostly to support physical oceanography programmes. It can operate to a depth of 2500 m and is being actively developed to meet oil exploration requirements.

Marconi

Marconi are drawing on their extensive torpedo experience to build a mine-countermeasures vehicle. Exactly how far this will be developed for commercial oil-field requirements is unclear.

Kongsberg-Simrad

Kongsberg-Simrad have developed a high precision untethered underwater vehicle (UUV) which can perform swath bathymetry and sidescan sonar

108 *High resolution site surveys*

functions without any direct cable connection to the survey vessel. This system can operate to a depth of 1000 m.

Japanese vehicles

Increasing Japanese interest in oil exploration has resulted in a number of attempts to produce a viable AUV, with a deep water diving capability. These systems may be summarised as follows.

The ALBAC prototype AUV is designed specifically for oceanographic measurement of the water column. The system consists of a cylindrical pressure hull and a pair of wings. The wings provide sufficient lifting force for the vehicle to move horizontally without consuming energy. ALBAC descends to its destination depth and measures water column parameters. On reaching the seabed a descent weight drops and the vehicle rises to the surface again, taking further water column measurements on the way. Diving is limited to about 300 m and the mission time is only about 30 min.

Twin-Burger is an AUV designed as a test-bed for software development. At present it can operate only in shallow water areas. The vehicle is powered by four 40 kW thrusters; two generate yawing movement, one is a vertical thruster and one is a side thruster. There are four degrees of freedom control modes. Vehicle positioning is achieved by dead reckoning navigation at the core of which is the AHRS (attitude and heading reference system). At present the vehicle can be used for camera observation of structures. No information on depth of operation or endurance is available.

The Manta-Ceresia system is a test-bed system for cruising sensors and has never been operated outside a test-tank environment. Based on test-tank cruising, the robot element learns to cruise by means of an artificial neural network.

The R-1 AUV robot consists of a main pressure vessel, a fore payload space and a propulsion system, contained in a torpedo shaped hull. At present the system carries a data logger, a colour video camera/recorder and a still camera. The propulsion system consists of a main thruster and two tunnel vertical thrusters. The motors are a DC brushless geared type contained in pressure housings. The vehicle is steered by control fins and elevators. Positioning uses an inertial navigation system and a doppler sonar. A collision avoidance sonar is also fitted. Vehicle position is telemetered to the ship using an super short baseline (SSBL) acoustic system. At present the vehicle can dive to about 400 m and can cruise for 20 h at 3.6 knots.

JAMSTEC are proposing to develop a new deep diving AUV based on the R-1 project. This vehicle will be able to dive to 3500 m and will have a range of 300 km at 3–4 knots. Propulsion may be by a prototype lithium ion fuel cell.

The UROV7K is a semi-autonomous vehicle which communicates through a 1 mm fibre optic cable. The vehicle descends under gravity. Near the seabed the vehicle releases ballast to achieve neutral buoyancy and manoeuvres by means of four thrusters. To surface, the remaining ballast is removed and the fibre optic cable is cut. After ascending, the vehicle is recovered by the

Analogue site survey systems 109

support ship. The vehicle is powered by a prototype lithium ion fuel cell. The UROV7K can descend to 400 m, has a range of 120 km and an endurance of 24 h at 2.3 knots.

The PTEROA-150 can dive to 2000 m but has a short duration of only 1–2 h. At present development is aimed at a longer duration vehicle capable of CTD (velocity, temperature, density) measurements, photographs and video.

The Aqua Explorer 2 (AE-2) is designed for sea floor and telecommunication inspection. The prototype was relatively deep diving and capable of going to 1000 m for 4 h. The AE-2 model has a hugely increased endurance at 24 h but can dive to only 500 m. The support vessel and the AUV are connected by two acoustic links.

Korean AUV development

In common with Japan, Korea is now a player in the AUV market. KRISO (Korean Research Institute of Ships and Ocean Engineering) have developed an AUV capable of diving to 200 m. It is known as VORAM (Vehicle for Ocean Research and Monitoring). It has no proper sensor system apart from a video camera. The endurance is limited to 4 h at 2.5 knots.

Daewoo, in collaboration with the Russian Institute for Marine Technology, have developed the OKPO 6000 vehicle, capable of diving to 6000 m. This vehicle was developed for bottom survey, mapping and seabed exploration. It descends by means of two sets of cast-iron ballast, one set being discharged when the target depth is reached and the other when the vehicle is ready to ascend. Propulsion is by four main propulsors installed in the stern. To date it has been used to map manganese nodules in the Pacific Ocean. It has also been used for inspection of sunken Russian nuclear submarines, one in the Sargasso sea at 5500 m and a second the *Komsonolets*, in the Norwegian Sea. The endurance of this vehicle is 10 h at 3 knots.

3.6 Sub-bottom profilers

Sub-bottom profilers were first developed in the 1950s and most are based on the magnetostrictive transducer principle. There is a wide range of systems which go under the general heading of sub-bottom profilers. They are variously known as profilers or pingers and occupy the next highest frequency range of seismic survey systems after the sidescan sonars and echosounders. Sub-bottom profilers operate in the kilohertz range, generally from 1 to 10 kHz. As systems they came into existence as another early adaptation of the echosounder transducer used to investigate sub-bottom sediments, usually in the region of the first 30 ms below the seabed. The pinger uses pulsed energy at a fixed frequency in a manner similar to the echosounder, the chief difference being that the output frequency is lower and the power output will be considerably boosted. This means that the penetration is considerably greater than the echosounder's but the resolution is not as good. Figure 3.14 shows a block schematic diagram of a profiler system.

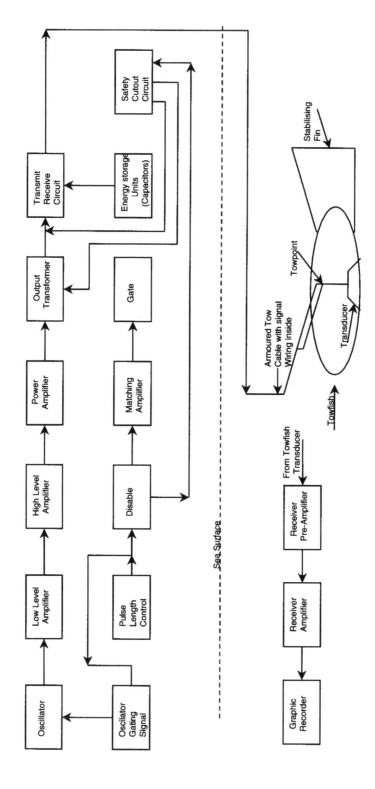

Figure 3.14 Typical profiler schematic.

Analogue site survey systems 111

The transducer is usually an electromagnetic coil. The transducer can be hull mounted or towed behind the survey vessel with the transducer mounted on a surface tow catamaran. Most pingers are mounted in a free-flooding 'fish' which is towed behind the survey vessel at a depth of 4–12 m below the seabed. Sub-tow or deep-tow pingers may be described as something approaching an industry standard.

Box 3.3
The mounting of profiler transducers
One contractor used a hull-mounted pinger system and found that the cavitation across the vessel bottom ruined the profiler records. After many attempts at improving the system the contractor eventually gave up and mounted the pinger transducer in a towfish. The improvement in record quality was enormous. Pinger systems are inherently prone to cavitation effects when hull mounted. A much better record is always obtained from fish-mounted systems.

A modern development is so-called 'chirp' profiling systems. In these systems a swept frequency acoustic output is used as opposed to the explosive wavelet of energy used by more traditional sources. Acoustically, for a given power output, a low frequency signal will penetrate into the earth further than a high frequency signal. The high frequency signal is also necessary because it is capable of resolving more closely spaced layers of sediment than the lower frequency signal. In practice, conventional systems use a compromise frequency selected to obtain the best tradeoff between penetration and resolution. The main advantage of a chirp-based system is the best resolution over a range of signals and both penetration and resolution are optimised. A chirp system can transmit large amounts of power into the water and maintain high resolution. In a conventional system high resolution means shorter pulses, with a resultant lowering of the transmitted power. The overall advantage of a chirp system is an approximate 40 dB effective gain.[13]

Most sub-bottom profilers consist of a power supply/control unit, an electromagnetic transducer which transmits and receives signals and a plotter. Sub-bottom profiler data will usually be recorded on tape. This is usually ordinary audio tape and any high fidelity audio recorder may be used. A time variant amplifier (TVG), swell filter or analogue stacker may all be used.

Typical pingers and profilers include the Edo-Western Microprofiler, the Ferranti 310c pinger pipeliner system, the Geoacoustics Geochirp pinger, the Geopulse ORE Model 5430A pinger, the ORE Model 140 profiler transceiver, the ORE 3000 pinger/pipeliner profiler and various Klein-Hydroscan systems. Just as echosounders and sidescan sonars can be integrated so sidescan sonars and pinger/profilers can be integrated. The Datasonics SIS-1000 is a good example. This system is a combined chirp sub-bottom profiling and sidescan sonar system. The advantage of an integrated sonar/profiler is that

112 *High resolution site surveys*

two systems can be placed in one towfish which can operate across a wide range of weather conditions with the minimum of interference from other seismic systems that may be run simultaneously. The SIS-1000 system uses a digital multiplexer for a single co-axial tow cable. Communication rates can be as high as 2 Mbit/s. The recorded data can be in a SEG-Y format or QMIPS.

In recent years a number of systems have been developed with echo-sounder multibeam capability, sub-bottom profilers and sidescan sonars combined in one towfish. This family of systems includes the Racal SeaMARC system and the Seafloor Survey International SYS system.

Sub-tow and deep-tow systems are usually fitted with heave compensation which compensates for fish motion as well as detected seabed position. Typical survey parameters might be as follows:

- Towing depth of fish: 3.0 m
- Record length: 125 ms
- Timing lines: 6.125 ms
- Firing rate: Four pops per second
- Record delay: 15.625 ms
- Operating frequency: 3.5 kHz (possibly dual frequency at 3.5 and 7.0 kHz)
- Power output: 8 kW
- Pulse length: 0.5 ms
- Bandpass filter: 5 kHz

Quality control

Quality control devolves around assessing the optimum frequency and filter settings for a particular survey area. In areas of hard clay, penetration is exceedingly difficult and a record that is little more than a bottom profile may be all that is possible. Sand channels and soft sediments may by contrast be mapped to a depth of 30 m or more. When penetration is poor then the lower transmit frequency will usually give the best results. In good penetration areas the frequency may be increased without loss of penetration but with improved resolution. Adjustments to the receiver begin with the choice of filter settings to cut out random noise and pass only those frequencies which carry coherent data. Gain settings are very important.

Figure 3.15 shows the effect of badly adjusted gain control. The central portion of the record shows what signals are actually received at the transducer and the remainder of the record shows the effect of poor gain control. Figure 3.16 shows an apparently good record with penetration down to three scale divisions. By asking that the gain be increased the consulting engineer onboard was able to show that penetration was obtainable down to four scale divisions and that weaker events from deeper horizons were being received but not recorded. This was due to the slope setting on the time variant gain being set too low. It is recommended that a swell filter should always be

Figure 3.15 Poor sub-bottom profiler record. Good reflections in 2 lost by very slight maladjustment to the overall gain.

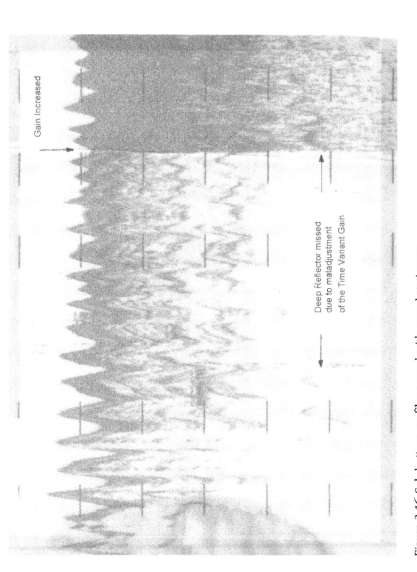

Figure 3.16 Sub-bottom profiler record with vessel stationary.

available and used in a sea state above 1.5–2.0 m. The layback and offset of the towfish should be accurately established before the start of survey.

Figure 3.17 shows a record section made using the same system in the same location after the system had been optimised. It should be noted that the excessive interference seen in Figure 3.15 is by no means so prominent in Figure 3.17.

The ideal tow position for any sub-tow or deep-tow system is from a point close to the centre of equilibrium of the vessel. This will minimise wave and swell noise seen on the records. If the system is to be used in higher sea states then the towfish must be run at a deeper depth.

3.7 Boomers and sparker profilers

There are two types of lower frequency profiler, usually known as boomers and sparkers. These systems can be operated as surface-tow, sub-tow or deep-tow systems. These systems are true seismic energy sources with a broadband frequency spectrum and a high peak intensity. In the case of some systems they require only the addition of a multi-trace streamer and digital recording system to turn them into digital seismic systems. In boomers and sparkers the energy discharge is from a bank of capacitors as indicated in Figure 3.18. The charged capacitor banks are discharged into an electromagnetic coil in the case of a boomer or into an electrode in the case of a sparker. Sparker systems are usually considered to give better penetration, while the boomer signal has a cleaner pulse. The pulse characteristics of such systems have been fairly extensively investigated over the years, both theoretically and experimentally. Contractors rarely have any hesitation in discussing the exact nature of their pulse systems. The electric to acoustic energy conversion is low in these systems, when compared to conventional sonar transducers, but it needs emphasising that boomer systems have the advantage of a wide frequency bandwidth and its concomitant: a high resolvability of seismic returns.[14]

In boomer systems a typical frequency spectrum might be in the range 200–8000 Hz. The seismic energy is created by electromagnetic means. The transducer is typically an insulated metal plate and a rubber diaphragm adjacent to a flat wound electrical coil. A short duration high power electrical pulse is then discharged from the sparker energy source into the coil. The resulting magnetic field explosively repels an aluminium metal plate. The plate motion is transferred to the water by means of a rubber diaphragm, generating a broadband acoustic pressure pulse. This device, suitably insulated, is placed in the water where the sudden movement of the plate causes a compressional wave to be set up in the water. An alternative system is a single electrode sparker system with the output power adjusted to give approximately the same frequency spectrum as for the boomer equivalent system. These systems are capable of obtaining seismic data up to 50–70 m below the sea floor.

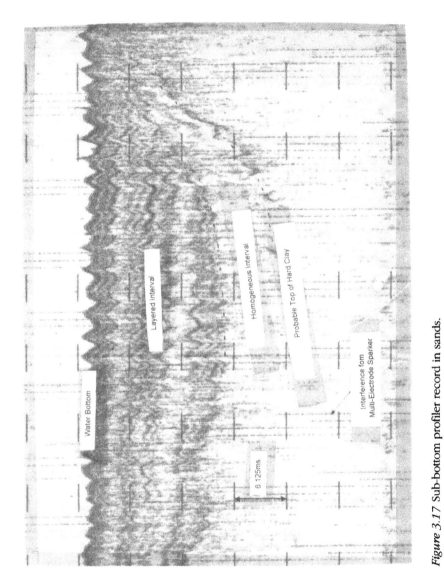

Figure 3.17 Sub-bottom profiler record in sands.

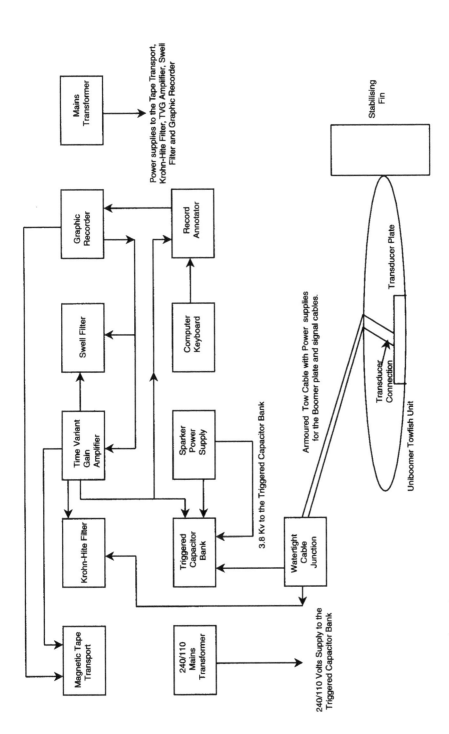

Figure 3.18 Sub-tow boomer system.

118 *High resolution site surveys*

The compression wave travels down into the sub-surface and reflections from the seabed and lower interfaces arrive at the surface where they are registered on a hydrophone. The signals are fed to a receiver which should have full band-pass/band-reject facilities. The signal is then filtered to retain only those frequencies which carry useful data. The receiver should also have overall gain and TVG controls to enable all the data present to be recorded.

The Huntec Sea Otter surface tow boomer is a typical catamaran-mounted system suitable for operation in good weather conditions. The catamaran supports the boomer plate and the boomer plate gives a high intensity acoustic pulse. The Boomer plate is an electromagnetically driven plane piston transducer. The power supply for this system is usually an EG and G 232 sparker discharge system. The output power is typically 300 J, with a firing rate of 350 ms, a record length of 130 ms and filter settings of 400–3000 Hz.

The EG and G model 240 surface towed boomer is another typical catamaran-mounted system. The acoustic pulse of this system does not have the strong cavitation or ringing pulse associated with many of the older boomer systems. Again this is a 300 J system. Power outputs of systems can be described in terms of power in joules and in terms of acoustic output. A 350 J system corresponds to 120 dB ref. 1 µbar at 1 m.

The Ferranti ORE Model 5813A is a third catamaran-mounted system with a bandwidth greater than 14 kHz at the 6 dB points. This system generates a full power output pulse of 50 µs duration, with full damping occurring within 150 µs from completion. The system can also be mounted in a sub-tow fish.

The Huntec deep-tow boomer was considered innovative in its time, though it is not much used today. This system used an electromagnetically driven plane piston transducer and emitted a single wide frequency band pulse. The capacitor discharge system was contained in the towfish and gave a power output of 500 J. The EG and G 240 deep-tow boomer is very similar.

The Nova Scotia Research Corporation deep-tow sparker system is typical of the alternative to a boomer. A capacitor discharge between electrode tips produces a frequency spectrum in the region of 200–10,000 Hz. Gardline surveys have developed their own deep-tow sparker system which may be considered a more reliable version of the Huntec and Nova Scotia systems, though it is unclear at the time of writing if it is still in service.

Typical boomer parameters might be as follows:

- Source level: 210 dB/µPa at 1.0 m depth (500 J)
  120 dB at 350 J, or perhaps 107 dB at 300 J
- Firing rate: Six pops per second
- Pulse length: 0.2 ms
- Frequency spectrum: 200–10,000 Hz
- Beamwidth: 30° to 40°
- Receiver sensitivity: −165 dB/V/µPa
- Bandwidth: 1 to 10 kHz
- Tow noise: 95 dB/µPa

Quality control considerations

In terms of quality control, insulation breakdown is a common cause of bad record quality. It is essential to check for leakage of the power cable to the boomer plate and for breakdown within the boomer plate itself. Since the energy source is a sparker discharge it is important to check the nature of the current pulse leaving the power source to ensure that the pulse is clean and sharp. For example, a double pulse will cause poor resolution on the records and this will severely limit the penetration of the shallow strata. Noise is always a problem with analogue survey systems and can be divided into a number of headings. Figure 3.19 shows a section of deep-tow sparker record with and without heave compensation. This record section also shows pipelines that are seen as parabolic signatures.[15]

Ship-generated noise

Ship-generated noise falls off inversely with distance from the survey vessel. Ship noise is not usually a problem with deep tow systems unless a short cable offset is used in shallow water.

Strumming of the tow cable

Cable strum can be substantially reduced by fairing the tow cable. There are a number of fairings available and some are better than others. Cable strum tends to be at low frequency, usually below the frequency spectrum of the acoustic sound source. The effect of cable strum can be further decreased by acoustic decoupling between the end of the cable and the fish tow-point.

Induced vibrations and resonances in the towfish

Towfish vibration and resonance are excited by the sound source firing. They are also initiated by heave and cable strum. Individual noise sources can be difficult to identify but, once done, the offending vibration can usually be damped by one means or another.

Vertical fish motion in response to heave on the tow cable

Vertical motion of the towfish due to heave can be reduced by using a high drag fairing such as faired fairing, but this can reduce the towing depth. A variety of different fairings may be used by a contractor for different survey conditions. Residual vertical motion can be reduced by electronic heave compensation. The vertical motion of the towfish should be continuously monitored by an accelerometer and by a pressure transducer. The accelerometer signal can then be filtered to approximate a double integration. The filter output is then proportional to vertical displacement over a short

120 *High resolution site surveys*

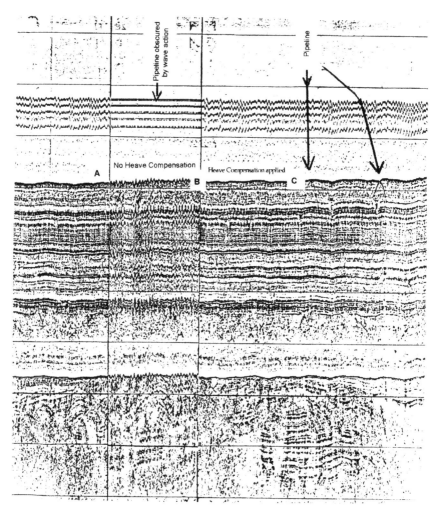

Figure 3.19 Sub-tow record using a deep-tow seismic profiler. The centre-left record section marked A–B shows a record with no heave compensation. Data in the form of a pipeline is obscured by wave effects on the sub-bottom data. The following section of record marked B–C shows the beneficial effect of using heave compensation. The remainder of the record shows a record section in which features such as a buried pipeline are indicated by parabolic signatures.

period of time. Longer period vertical displacements are obtained from a pressure transducer. These two signals are then used to advance or retard the trigger point of the source relative to the graphic recorder. The end result is a seabed profile corrected for fish motion. This is shown in Figure 3.19 which clearly indicates the advantage of heave compensation.

Analogue site survey systems 121

Pitching and rolling of the towfish arising from heave

Pitching of the towfish in response to heave is reduced by minimising the heave. This is done through decoupling and improved towfish design. On most heave-compensated systems a vertical motion of about ±3 m results in a pitch amplitude of ±3°.

Electrical noise and crosstalk

Electrical noise has been steadily reduced over the years and is not usually a problem on modern systems.

Ambient sea noise

Ambient sea noise is dependent on sea state, wind speed/direction and the sea floor type. Working in shelf seas, ambient sea noise should be more or less constant with water depth. Most systems are engineered so that the transducer plate acts as a reflector, shielding the internal hydrophone from downward arriving energy and reducing the effective noise aperture. Most sources state that an internal hydrophone is about 3 dB quieter than an external hydrophone due to this effect.[16]

Hydrophones for boomers

Most boomers and single electrode sparkers use a single hydrophone as the receiver element, contained within the towfish, which gives the advantages described above. Surface-tow systems are not limited to the single point hydrophone and may use a mini-streamer. A typical mini-streamer, such as an EG and G 265 mini-streamer, consists of eight hydrophones spread over a distance of 15 ft. Each hydrophone has typically a sensitivity of 5 µV/µbar. A more modern mini-streamer is the HNB eight-element mini-streamer. Ideally the hydrophone mini-streamer and boomer catamaran should be towed symmetrically either side of the vessel wake.

Filtering and display

Since boomers and sparkers are not discrete frequency devices the filtering of received signals is very important. Figure 3.20 shows a deep-tow sparker record and tests conducted with different filters. To set up the filters the main noise frequencies such as ship noise, cable noise, sea noise and noise from other geophysical systems must be reduced to a minimum. This done, the frequency band that contains the reflected signals must be found. Most contractors adopt a trial-and-error approach, adjusting the filter settings until the best record is obtained. The best-looking record is usually considered to be the one with the least noise. This is not necessarily so and records with

122 *High resolution site surveys*

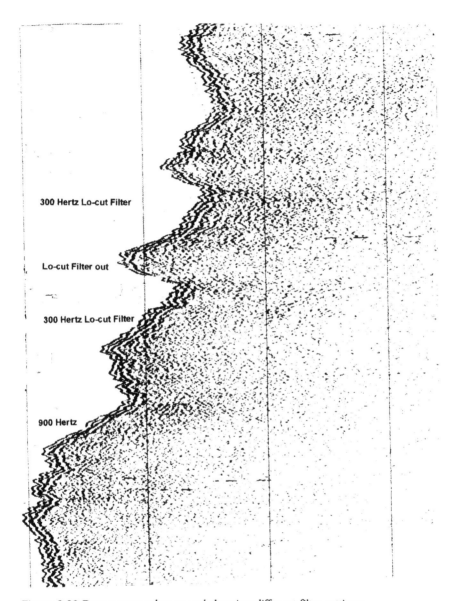

Figure 3.20 Deep-tow sparker record showing different filter settings.

some noise may actually contain more geophysical data than a record that has been filtered to eliminate all noise. Another bad practice is to increase the threshold below which no signals are written, that is to write on the record only the strongest signals. Undoubtedly the record looks prettier without noise but much useful data would have been lost. Legible noisy records with the maximum of data are preferable to clean noise-free records with only the strongest data showing.

Data from boomer and sparker systems is often recorded on an analogue recording system such as a Delph-1. This system is based on a 386 IBM compatible computer. Raw unfiltered data is usually recorded on tape and this raw data can be reprocessed at a later date, with different filter options if so required. Boomer and sparker data can also be digitised and multiple removal is possible. Amplitude manipulation, improvement in the signal-to-noise ratio and removal of wave or tow depth effects can also be carried out.

In conclusion, the boomer is a shallow investigation high resolution tool and its use is similar to the pinger/profiler systems. Being a lower frequency broad pulse system the resolution of the boomer is less than that of the pinger/profiler systems but the penetration is greater. Most site surveys use either a boomer or a pinger/profiler.

3.8 Multi-electrode sparkers

The multi-electrode sparker (MES) is a shallow investigation tool capable of penetration to a depth of 100–250 m with reasonable resolution. It is particularly suited to the location of geological problems likely to be found during the driving of conductor pipes. Boulders, fluid sands, gas pockets, shallow fault pockets, channels and steep dip interfaces, etc. can all be delineated with a multi-electrode sparker.

Figures 3.21 and 3.22 show the physical form of a multi-electrode sparker. The tow sled typically consists of 144 sharp wires protruding from an insulated central electrode. The seismic energy source is created by a high voltage electrical discharge into the water through the sharp wires (tips). The electrical discharge vaporises the water, causing a small explosion. A compressional wave travels out through the water in all directions. Reflections from various horizons arrive back at the hydrophone which is towed at the surface of the water. The signals are filtered, amplified and fed to a graphic recorder. The hydrophone, recorder and power sources are usually those used with the boomer and single tip sparker. One-hundred-tip and thirty-tip multi-electrode sparkers are occasionally used and at least one contractor has a catamaran-mounted MES system.

The Geo-Resources multi-electrode sparker is mounted on a towed catamaran sled and consists of 180 sharp wires protruding from an insulated central electrode. In this system there are 100 'fine' tips and 80 'fat' tips. The two sets of tips have differing power levels through them, about 300 J for the fine tips and 400 J for the fat tips, though this is variable. The frequency spectrum produced can be much more linear than in older systems. The Geo-Resources MES is unusual in that the tips are negative and the earth is positive. This causes much less tip wear than older systems. After firing (described above) reflections from various horizons arrive back at the single trace mini-streamer which is towed at the surface of the water. The signals are filtered,

124 *High resolution site surveys*

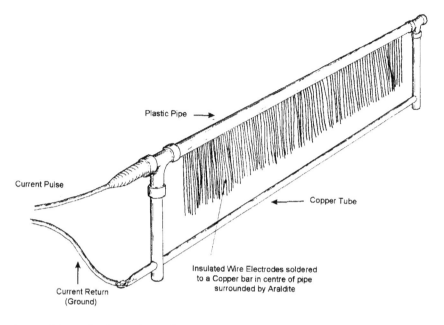

Figure 3.21 Multi-electrode sparker type 'A'.

amplified and fed to a graphic recorder. The hydrophone, recorder and power sources are usually those used with the boomer and single tip sparker. Figure 3.23 shows a schematic of a complete MES system.

The MES operates with an energy output between 200 and 1500 J, depending on conditions, water depth, hardness of the seabed, penetration into shallow strata, etc.

Box 3.4
New age multi-electrode sparkers

These instruments have been unpopular for many years but the Geo-Resources multi-electrode sparkers are modern instruments that overcome all the inherent failings of the older systems which date from the 1960s and 1970s. At the time of going to press the Geo-Spark DT1000 is entering service. This is a deep-tow system and represents the first attempt at building a deep-tow multi-electrode sparker. A deep-water sparker has also been developed with 800 tips, operating at 4.5 kJ. In addition there is a fresh-water sparker for harbour and estuarine use. The prospect of all-weather deep-tow multi-electrode sparker data should represent a major breakthrough in high resolution shallow seismic profiling.

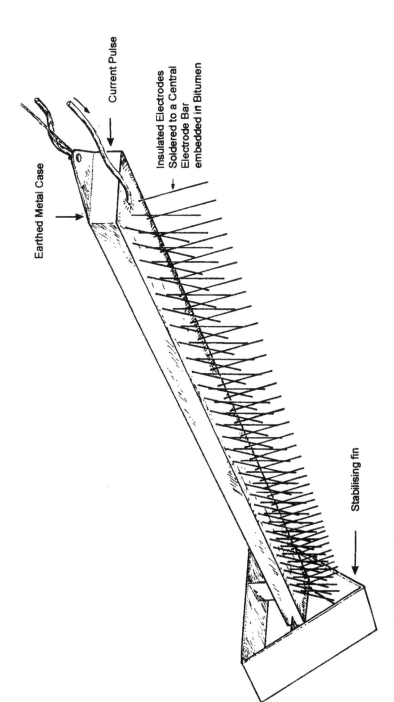

Figure 3.22 Multi-electrode sparker type 'B'.

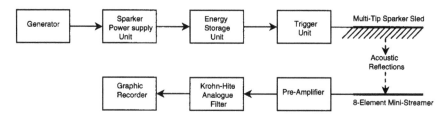

Figure 3.23 Multi-electrode sparker transmission and reception.

MES data is normally received on a short multi-element mini-streamer of the type already described. An EG and G 265 nine-element mini-streamer is typical and suitable for the higher frequency end of surface applications. The Design Products mini-streamer appears to be a more or less exact copy of the older EG and G mini-streamer. There are other EG and G mini-streamers available such as the 263 which is suitable for most sparker and multi-electrode sparker work. Alternative mini-streamer systems include the Geomechanique 30 m, two-channel summed array which is very sensitive to a wide range of frequencies. This two-channel system has a signal-to-noise advantage of 1.414 ($\sqrt{2}$). With a high resolution seismic source and receiver the systems are usually towed close to the sea surface. If the bandwidth of the seismic pulse is considered to be 300–800 Hz and a dominant frequency of 500 Hz is taken as the median, then the mini-streamer and source should be towed at a depth of 1.5 m.

Towing geometry is very important for multi-electrode sparker operations (see Figure 3.24). The sled and mini-streamer should be towed symmetrically either side of the vessel wake. This shields the hydrophone from the direct pulse and lower order outputs of the sparker source and helps to eliminate a great deal of clutter from the records. The seismic echoes received at the hydrophone are usually 1/5 to 1/10 or less of the amplitude of the direct pulse from the MES. If lower order acoustic radiation were received from the acoustic energy source its effects on the hydrophone would be comparable to the seismic echoes if no shielding action from the wake were achieved. The first and third bottom multiples are greatly reduced in amplitude because of the scattering effects of the propeller wake. Vessel speed is important in that the wake increases with speed and provides a better shield and scatterer. A boat speed of 3–6 knots is optimum.

Typical recording parameters might be as follows:

- Power supply: 1 kJ
- Record length: 500 ms
- Bandpass filters: 600–3000 Hz
- Firing rate: Two pops per second
- Source depth: 1 m
- Streamer depth: 1 m or less (possibly surface tow)

Analogue site survey systems 127

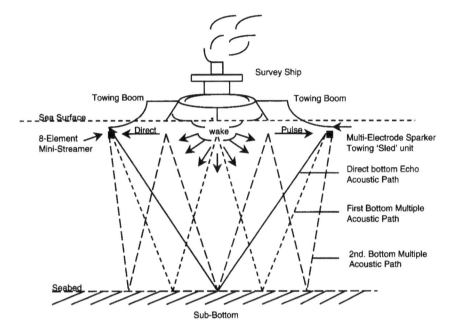

Figure 3.24 The shielding effect of vessel wake illustrated. This diagram shows the effect of towing symmetrically on opposite sides of the vessel wake. Symmetrical towing shields the mini-streamer from the direct pulse and lower order outputs from the sparker unit and this eliminates a good deal of clutter from the records. The seismic echoes as read at the mini-streamer are usually one-fifth to one-tenth of the main pulse, with effects at the mini-streamer comparable to seismic echoes if no shielding action from the vessel wake were achieved. The first and third bottom multiples and to some extent the second multiple are greatly reduced in amplitude because of the scattering effect of the survey vessel propeller wake. Vessel speed is important in that the wake increases with speed and provides a better shield and scatterer. Survey vessel speeds between 3 and 6 knots are optimum.

Quality control

Quality control considerations start with the condition and method of deployment of the electrode unit. Insulation breakdown and corrosion inside the cable splice to the electrode unit is the most common cause of system failure. Most contractors have a tendency not to resplice the main electrical connections before the start of survey, yet time and again this has to be done during the survey. A poor record due to difficult bottom conditions looks similar to a poor record due to electrode leakage, poor deployment or a low sensitivity hydrophone and poor definition of the discharge current, etc. Claims that the system is working properly based on the visual inspection of records can only be a positive test if the previous records shot at the same

128 *High resolution site surveys*

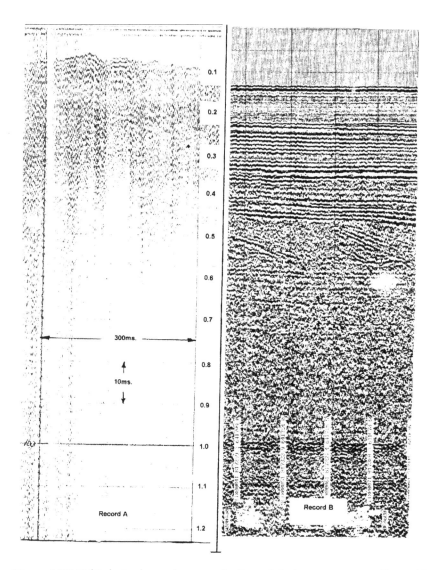

Figure 3.25 Multi-electrode sparker records. Record A shows a poor quality multi-electrode sparker record. All too typical of MES records without proper QC supervision. Record B shows a good quality multi-electrode sparker record. Typical of MES records with proper QC supervision.

location are available for comparison. Figure 3.25 shows poor and good sections of multi-electrode sparker records. These illustrate just how bad the MES record can actually be, without proper setting up and adjustment. Figure 3.26 shows a good quality MES record with excellent data down to the first multiple at 175 ms (130 m). A valley feature shows prominently.

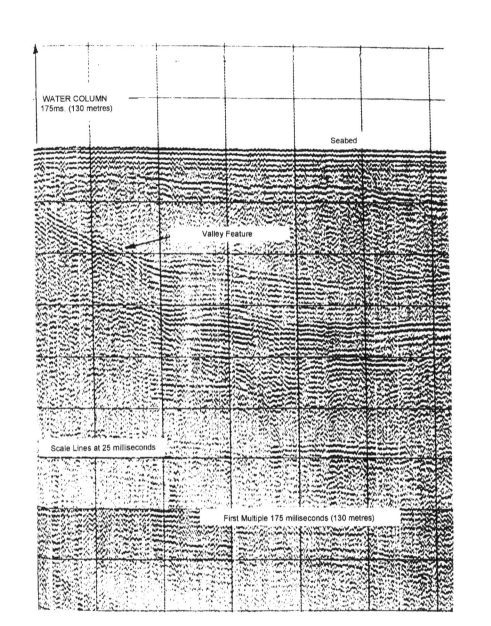

Figure 3.26 Good quality MES record. This record shows excellent analogue data down to the first multiple at 175 ms (130 m). A valley feature shows clearly. With digital recording the multiple could be removed, further extending the depth to which data can be recovered.

130 *High resolution site surveys*

The MES is a polarity sensitive tool and the sparker polarity needs to be checked after deployment. This system cannot give good results in poor weather. Best results are obtained in a sea state of three or less. This system is not greatly used today because it cannot be used as part of an integrated package of systems. The interference of the MES with other systems ensures that it has to be used as a stand-alone system. There is no particular reason why the MES should not be used with a multi-trace streamer and digital recording system, other than convention.

3.9 Analogue recorders and filters

At one time, wet paper sepia recorders were in widespread use by the site survey industry. In the main this type of analogue recorder is obsolete but a few remain in use, chiefly for sidescan sonar record presentation. These recorders produced wet paper records that had a short shelf life but a good visual dynamic range for interpretation immediately after data acquisition. The Gifft wet paper recorder was a typical example of this type of recorder. In common with most wet paper recorders the record marking electrodes consisted of a grounded blade which oscillated slowly to distribute normal erosion over the entire blade. There were three contacts mounted on an endless metal blade which ran on two pulleys and was driven at a precise speed by a sweep mechanism. The record was written by a voltage that caused darkening of the paper, in proportion to current flow. In common with most recorders the Gifft unit was crystal controlled and a 360 line bar pattern printed directly from the crystal frequency standard was the best overall quality control check on the system.

Probably the best known recorders in use are the EPC graphic recorders, long regarded as an industry standard. These recorders are still in widespread use for displaying sparker data, boomer data and practically any seismic survey system in use. The recorder stylus is a moving belt that starts recording when the stylus passes a datum point. As the stylus moves across the chart the recorder initiates the sparker, boomer or pinger firing sequence and data appears in its correct time sequence as the stylus progresses across the recorder paper. In terms of quality control the condition of the stylus is critical to record presentation and causes a great deal of difficulty when not properly set up.

The Benson–Schlumberger dry paper record was a big advance in recorder technology but not much used in the site survey industry because it was expensive and the recorder paper was very expensive indeed. In recent years the OYO Geospace GS 622 thermal plotter has superseded most of the older recorders. This unit is comparatively bulky but is reasonably cheap and easy to set up. Most modern site survey operations use a combination of EPC graphic recorders and OYO Geospace recorders. In so far as industry standards exist, most site survey instrument rooms include at least two EPC recorders and two OYO Geospace recorders.

Digital recorders are now in widespread use for analogue instrumentation recording. Data from echosounders, sidescan sonars, pingers, profilers, sparkers and boomers is displayed on the analogue recorders described and digitised by the digital system in use. The Delph seismic recording system is a good example of this type of instrument. Sample rates can be as fast as 20 μs and up to twenty-four channels can be inputted. Two channels of data can be acquired and processed asynchronously according to two independent sets of parameters, including two different sample rates, provided one is a multiple of the other. Bandpass filters, swell filters and automatic gain control (AGC) can also be applied. Spectral analysis, signature deconvolution, predictive deconvolution and time variant filtering are also available. Common depth point (CDP) processing is possible with at least six channels in use, thus improving the signal-to-noise ratio. The Delph system uses a 233 MHz Pentium processor as the host processor and has 64 MB RAM; 8 or 16 GB of hard drive is available.

Analogue filters are still used to filter boomer and sparker data. They provide the correct bandpass filtering prior to printing on a graphic recorder. A typical analogue variable bandpass filter is the Krohn–Hite filter used by the site survey industry since the early 1970s. This unit has a maximum flat response derived from a fourth order Butterworth function. The bandpass is obtained by setting the frequency dials to the desired low and high frequency settings and establishing the maximum flat response. There are other analogue filters in use but the Krohn–Hite is typical.

Swell filtering is used to remove static caused by wave action or swells. Swell filtering increases the operational weather window across which survey data may be acquired but signal-to-noise ratio will always degrade in bad weather. Heave compensation is useful for sub-tow systems and again removes static caused by wave action, swell or towfish height adjustments. Such systems are based upon accelerometers or pressure transducers.

3.10 Seismic refraction survey

Seismic refraction techniques were relatively common in the 1960s and 1970s but were not much used in the 1980s and early 1990s. There has been a resurgence of interest in seismic refraction with the development of the Fugro Gambas system. In this system seismic refraction is integrated with seismic reflection to produce an enhanced shallow sub-sea dataset. Velocity control is much more precise with both reflection and refraction data. The Gambas system consists of a sledge, towed on the sea floor, carrying a hydraulic power pack, a data acquisition and transmission unit and a seismic source. A hydrophone streamer is towed behind the sledge. A stop-and-go motion device enables the sledge to remain stationary and the streamer to be in full contact with the sea floor, during shooting and recording sequences, while the survey vessel proceeds at survey speed. A sensor package controls the position and motion parameters of the sledge. Processing and preliminary data interpretation are performed onboard the survey vessel, with the recorded signals demultiplexed and configured in SEG Y format.

Other older methods of acquiring seismic refraction data are described here for historical completeness as much as anything else. The seismic source was usually an airgun or sparker with a power output of at least 5 bar m, the same level as for a digital seismic survey. A sonobuoy was used as a receiver and transmitter. The hydrophone was attached to the sonobuoy, usually on an elastic suspension designed to dampen out sea wave action. The buoy contained an amplifier and radio transmitter which transmitted seismic signals to the shipboard receiver. Acoustic signals were displayed either on a multi-channel oscillograph recorder or on a graphic recorder as well as being recorded on magnetic tape for subsequent processing. The shot instant was recorded and from the display it was possible to detect, as well as the first arrival signal, a later strong high frequency signal. This was the acoustic impulse which had travelled directly through the seawater from source to hydrophone.

Quality control

Quality control on the refraction method is chiefly concerned with measuring interval times which are crucial to the acquisition of valid data:

1. Shot instant to the first arrival;
2. Shot interval to direct wave arrival.

The second time interval divided by the velocity of sound through seawater gives the distance between the shotpoint and the sonobuoy. Such a measurement is particularly important if the sonobuoy is free floating because the position of the buoy will vary over the duration of the profile acquisition period.

It should be understood that refraction techniques as applied to site surveys were only ever used for velocity data.

3.11 Site survey discussion

As a conclusion to this discussion on analogue site survey systems an analysis is needed that covers the use of the systems as a complete package. Site surveys tend to produce a huge amount of wasted time and records are often of poor quality. Since three or four systems are used simultaneously all must work at the first deployment of survey equipment on site. It is very common that one system fails or takes a great deal of time to be made operational. Generally speaking, the survey should not start until all the analogue systems can be run as a package. If a large number of reshoot lines occur due to running single system lines then the contractor should pay for the second and successive line attempts. Pre-survey checks can reduce the amount of lost time at the start of survey. Tests in port (static tests) involve deploying the equipment in harbour while the vessel is tied up alongside the quay. Such tests establish that the basic electronics of the systems is working; however, no indication of the expected record quality or reliability on site can be obtained from such tests. Leakage on cable systems can be checked in port and cable splices can be remade. Digital instruments can be checked and the tests analysed by a Micromax or Promax system. Hydrophone sensitivity can be tested and current pulse can be measured on sparker systems. Gun and sparker signatures, electronic noise levels, etc. can be checked in port. Most contracts include a flat rate mobilisation charge and this should include a complete check on all survey systems before the vessel leaves port.

Most site survey contracts are run on a daily rate basis and the client representative must strike a balance between the quality of records and the time spent on site overhauling and testing equipment.

References

1. Atlas-Deso 20 Manual.
2. Geo-Acoustics data sheets covering dual frequency sidescan sonar and the Geoprocessor processing sequence.
3. 'Comparison of sidescan sonar types', *Hydro Magazine*, May/June 1998.
4. Clasper, P. (1996) 'Aspects of quality control in swath bathymetry for pipeline route surveys', *Hydrographic Journal*, 79, January.
   Miller, J.E., Hughes-Clarke, J. and Paterson, J. (1997) 'How effectively have you covered your bottom', *Hydrographic Journal*, January.
5. Hutchins, R.W. (1979), 'Development of new geophysical methods for site investigation', *Offshore Site Investigation*, proceedings of a conference held in London, March 1979.
6. Simrad EM 300, New Multibeam Echosounder, document, supplied by Simrad, 2 pp.; also document on the Simrad Neptune system, supplied by Simrad, 5 pp.
7. ISIS-100 information pack.
8. Geo-Acoustics data sheet.
9. *Simrad Post Processing of Bathymetry*, manual, supplied by Simrad, 7 pp.
10. Information pack, supplied by Bob Barton Associates.
11. Information pack, supplied by Bob Barton Associates.
12. 'Hugin Underwater Vehicle (UUV)', *Hydro Magazine*, May/June 1998. (This article also describes the Autosub and Martin 200 systems.)
13. *An Introduction to Geochirp*, document, supplied by Geo-Acoustics, 10 pp.
14. Hutchins, R.W. 'Huntec deep tow system', document, Huntec '70 Ltd., 20 pp. (This document has the deep-tow record shown in Figure 3.14.)
15. Ibid.
16. Ibid.

4 Non-seismic site survey techniques

4.1 Gravity corers and seabed sampling

The most common form of seabed corer is the UMEL Sargent gravity corer. Another gravity corer in fairly widespread use is the Kullenberg gravity corer. A typical example of the former unit, shown in Figure 4.1,[1] is 58 in. long. The bottom section is a 3 ft or 6 ft barrel with a core cutter at the bottom. The body of the corer has a weight locking clamp at the top and a maximum total of twelve 75 lb lead weights can be added to the body. The barrel has a plastic tube inserted with the core cutter at the bottom. The corer is usually deployed from a chute and suspended about 15 ft above the sea floor. A trip mechanism is initiated and the corer falls to the sea floor. The core cutter penetrates the sea floor and the weight of the corer is hopefully sufficient to drive the corer deep into the sea floor, producing a sample which is forced up into a plastic tube inserted inside the core barrel. The plastic tube is removed with the core sample in place. Other corers in use include the Benthos gravity corer. A grab bucket is sometimes used in place of the gravity corer. A van Steen grab bucket is typical.

Box 4.1
Coring and smelling
When the gas fields were being surveyed in the southern North Sea, local fishermen claimed that they had, over many years, frequently smelt gas, an assertion largely ignored by the geophysicists of the time. The fishermen were entirely correct and seabed cores should be smelt when they are analysed onboard the survey vessel. The 'rotten egg' smell of hydrogen sulphide is a very good indication of gas in the core and gas further down. Sidescan sonar records sometimes reveal 'pockmarks', gas bubbles that have come to the seabed. Again, local fishermen often claim to have smelt gas in pockmarked areas. Smell can be a useful adjunct to core analysis.

Coring is usually done at the end of the survey. Often it is a fairly hazardous occupation with the corer swinging around on the deck of a small survey vessel, heaving and pitching in a seaway.

Non-seismic site survey techniques 135

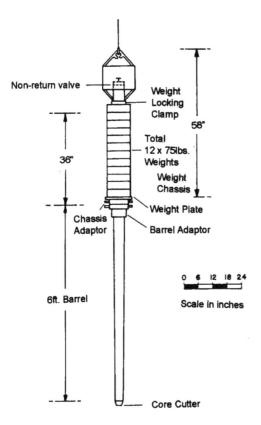

Figure 4.1 UMEL Sargent gravity corer.

Most site surveys require three, four or five cores. Usually one is taken at the centre of the survey grid and others are taken either on a regular pattern or where there are marked changes in the graininess of the sidescan sonar records. It is much better to determine the core positions on site and establish the nature of the sea floor. A particular graininess on the sidescan records can then be determined to be sand or gravel. Another characteristic graininess might turn out to be silts and clay. More exacting analysis may be performed to determine the exact grain size of the sample. Intelligent integration of the sidescan records with the coring programme can produce a greatly enhanced survey data set.

4.2 Shearvane penetrometers

The penetrometer consists of a sounding rod driven down by a hammer. The penetration resistance of a particular area of the seabed is calculated by knowing the number of uniform hammer blows to obtain a particular level of

136 *High resolution site surveys*

penetration. Such information is valuable for assessing sea floor jacking conditions and anchor holding ability.

The sounding rod is driven down by the hammer, the height of fall being 50 cm. During penetration the rod is turned two turns every 20 cm.

It is rare for a seabed penetrometer to be used and most surveys use a small handheld shear vane unit for testing cores. Torvane and shear readings are taken on the core samples.[2]

4.3 Vibrocorers

The basic vibrocorer system consists of a modular frame of tubular construction and a twin linear vibrator motor housed in a pressure vessel. The unit is lowered to the sea floor and the vessel must be held on location while the sample is taken. The core cutter is vibrated continuously for about 20 min, during which time it sinks deeply into the sub-surface strata. Survey vessels that conduct this type of survey must be fitted with a bow thruster and variable-pitch propellers. Holding the vessel up to location is the most important quality control parameter. As with the gravity corer the sample is retained in a plastic liner tube. Maximum retention of sample is ensured by the operation of a piston in the liner tube and a catcher in the cutting head. Figure 4.2 shows an Aimers McLean vibratory coring system. A rather similar system is the Senkovitch vibrocorer.[3]

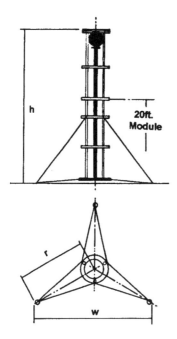

Figure 4.2 Aimers McLean vibratory coring system.

4.4 Underwater photography

It is sometimes necessary to carry out a visual inspection of a platform location or of wellheads that are to be recommissioned. The most common method is to use a remote-operated vehicle (ROV) installed onboard a site survey vessel. A typical system is the Seaeye Surveyor. The underwater unit is electrically operated, 1.5 m in length, 1 m high and 1 m wide. The weight of the system is 300 kg. The supporting framework is made of plastic, which assists buoyancy as do two flotation units. Such a system can work at water depths up to 1000 ft (300 m). A typical ROV has eight thrusters. Four of these form the main axial thrusters which provide the forward and reverse movement. They also provide the clockwise and counter-clockwise movements. Of the remaining four thrusters, two are for lateral movements and two for vertical movements. Each thruster gives 10 kg of thrust.[4]

The ROV is fitted with four cameras, three of which can be displayed at any one time. Typically, one camera will be full colour, one will be black and white, one will be a low intensity light camera and the last camera may in fact be a very short range scanned sonar unit, operating at an acoustic frequency 455 kHz. The colour unit is sensitive to 0.5 lux and the low light unit is sensitive to 0.001 lux, making it ideal for low light areas.

Box 4.2
Searching for wellheads with an ROV

It is sometimes the case that wellheads are to be reactivated after a lengthy period of deactivation. It may be the case that a site survey is performed before reactivation. An ROV may then be used as part of the site survey package. Using ROVs from conventional survey vessels produces a number of problems that need to be addressed. Vessels without an effective bow thruster should not be used for ROV work. The survey vessel needs to be stationary, held up into wind and sea for up to an hour at a time, and this is virtually impossible without an effective bow thruster. ROVs are delicate instruments and need to be deployed with care, not left swinging from a crane and banging against the side of the survey ship during deployment and recovery. The exact method of deployment and recovery should be addressed when the choice of survey vessel is made. Too often the survey team are left to sort out these problems on site.

4.5 Current meters

Current meters are often deployed from site survey vessels or left anchored at specific locations and recovered later. Current meter data is widely used in riser design and rig site assessment. Most current meters are sealed instrument

138 *High resolution site surveys*

packages capable of operating in water depths of up to 2500 m. Most systems in general use measure five parameters. These are, current speed, current direction, water temperature, water pressure and water conductivity.

Typically an electromechanical analogue-to-digital encoder samples and converts the different parameters to binary signals which are recorded on magnetic tape or perhaps on a 3.5 in. mini-floppy disk. The data is then played back after the meter has been recovered.

Typical current meters are the Anderra system and the Greystoke system.[5] A current meter deployment is shown in Figure 4.3.

4.6 Engineering tests and cone penetration

A particular type of site survey is sometimes performed with a small drillship or survey ship to make cone penetration tests and plate loading tests on a platform site, in order to establish the load bearing characteristics of the seabed. The information obtained will help to decide on the locations of shallow boreholes. Shallow drilling and coring can follow, with samples tested onboard before being sealed and despatched to soils engineering laboratories onshore.

In the cone penetration test a steel cone is loaded and rested on the sea floor. The penetration of the unit gives a measure of the resistance of the soil to piles, etc. The plate loading test involves loading a steel plate resting on the seabed until the soil begins to yield (yieldpoint). This gives indications of the load bearing capabilities of the subsoil. Sleeve and point resistances on the cone are measured, in conjunction with pore pressures. From this, information can be obtained on soil lithologies and an estimate can be made of the shear strengths of clays and the relative densities of sands. A modern system should be able to measure sediment type, stratification, density and shear strength in water depths up to about 2000 m.

Systems that perform this type of work are often referred to as cone penetrometer (CPT) systems that perform piezocone penetrometer tests (PCPTs). These systems usually require a heavy lift capability onboard the survey vessel, but lightweight systems are now available. PCPTs are performed at intervals which vary according to the nature of the seabed. If the stratification and soil properties are constant and uniform over relatively large distances the CPT spacing can be several kilometres. In variable conditions the CPT spacing may need to be much closer.

Seabed resistivity comes under the general heading of engineering tests and produces data used to predict soil lateral variability. Such data is sometimes required when shallow soil variation will have a deleterious effect on a rig moving onto a particular location.

4.7 Remote-operated vehicles

In the context of site surveys remote-operated vehicles (ROVs) are used mainly for pipeline inspection work. Since they are not specifically site survey

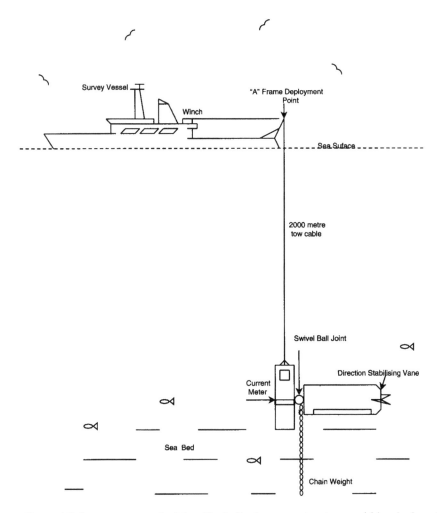

Figure 4.3 Current meter principles. Typically the current meter would be deployed from the 'A' frame aft. It is then free fallen to a few metres above the seabed and raised to the surface stopping for 10 min or so at each specified station. After each 'dip' the data tape is replayed and the data interpolated from graphs. More modern systems telemeter the data to the sea surface for recording immediately on a computer hard drive. The other method of obtaining current meter data is to drop the current meter onto the seabed and to recover it at the end of the survey by means of an acoustic release system. (Note: The current meter and survey vessel are not drawn to the same scale.)

instruments they have been included in this chapter. They could just as well have been included after the section on automated underwater vehicles. Section 4.4 of this chapter has already mentioned the Seaeye Surveyor, widely used for underwater photography. Typical ROVs are the Macartney Focus

140 *High resolution site surveys*

ROTV vehicles or work class ROVs. Work class ROVs are used for a variety of tasks, including inspection, drill support and survey work. In deep water below 1000 m ROVs are the principal means of obtaining route and field survey data. As an example, a contractor recently fitted an ROV with a multibeam echosounder, sidescan sonar and sub-bottom profiler to conduct a survey in 2200 m of water. ROVs are also used for pipeline video surveys (see Chapter 1.12). At the time of writing it is estimated that about 10 per cent of available ROVs are used for survey operations, representing about fifty vehicles. Four ROV systems are reviewed here.

Focus ROTV

The Focus 400 is a remotely operated towed vehicle (ROTV) designed for route surveys. A larger model, the Focus 1500, can carry a laser linescan camera. The vehicle has a 'box kite' body with parallel vertical and horizontal control surfaces for manoeuvring and positioning. The Focus 400 can operate to a depth of 400 m at a speed of 5 knots. It can carry tracking systems, a motion reference unit, dual Mesotech altimeters, a Digiquartz precision depth sensor, a pinger and a flasher. It can also carry a dual frequency sidescan sonar, obstacle avoidance sonar, scanning sonar, multibeam sonar and a subbottom profiler. For pipeline work a video camera with focus, and a still camera can be carried. Other options include a magnetometer, a laser linescan camera, a sound velocity probe, a CTD probe and a sampler.

The Focus 1500 can operate to a water depth of 1500 m and may be fitted with enhanced sensors such as a Seabat 8101 multibeam echosounder, a Northrop Grumman SM 2000 laser linescan system, a Klein 2000 digital sidescan sonar and a 3.5 kHz sub-bottom profiler. A Geometrics 880 caesium magnetometer is also an option.

Both systems are towed by a fibre optic towcable. This is constructed from seven light copper-armoured multimode fibres with an overall contrahelical steel armour. The copper armour on the fibres is used to transmit power to the vehicle.

Racal SeaMARC

The Racal SeaMARC system is a towed, bilateral sonar imaging system capable of simultaneous swath bathymetry and sidescan sonar operation. This system is designed specifically for route surveys. A deep-tow sonar operates at 50 kHz and is designed for plough-route surveys to 2000 m. The sonar can also undertake 'chirp' profiling. SeaMARC will provide 0.5 m resolution at a swath width of 1000 m. There are 2048 bathymetric and backscatter image samples across the swath for each transmission and reception cycle. The system uses a fibre optic towcable. Further developments may include multibeam side-looking sonar and synthetic aperture sonar.

Non-seismic site survey techniques 141

Stealth 3000 ROV

The Stealth 3000 ROV is manufactured by Hitec AS in Norway, for Seateam, now DSND Oceantech. It was designed for an operational depth of 3000 m and a survey speed of 5 knots. The permanent survey sensors include an Anschutz gyro, a Paroscientific Digiquartz depth meter, a CTD unit, motion reference sensors, Doppler velocity sensors and obstacle avoidance sonar. Optional extras include a multibeam echosounder, a pipe tracker, a cable tracker, sidescan sonar, responder/transponder, cross profiler, freespan detector and a CP probe. Three video systems can be fitted and six cameras.

Sea-floor Survey Systems SYS100

The SYS100 is a high resolution vector sidescan sonar. This is a 100 kHz system that operates to a water depth of 1000 m. It is fitted with a CHIRP sub-bottom profiler.

4.8 Gravity survey

Gravity methods represented the first geophysical attempt to prospect for oil by means of scientific instrumentation. As far back as 1876, C.W. Siemens constructed a sea-going gravimeter intended for water depth measurement. This was not, in fact, successful. During the 1920s and 1930s the Dutch geophysicist F.A. Vening Meinesz carried out a long series of experiments using a three-pendulum apparatus mounted in a submarine.[6]

Gravity survey is not normally a part of the standard site survey package. In the 1970s, gravity survey was an almost inevitable part of 2D surveys, with gravity platforms permanently installed onboard the survey ship. There is no particular reason why gravity platforms should not be installed onboard site survey vessels, so an account of the gravity method is given here.[7]

The most widely known gravity meter is the Lacoste and Romberg air–sea gravity meter, in use since 1967. This system consists of a highly overdamped, spring type gravity sensor mounted on a gyro-stabilised platform with the associated electronics for obtaining gravity readings and for recording the results on stripchart and magnetic tape records. The overall gravity range without resetting is 12,000 milligals with a static accuracy of better than 0.01 milligals. One gal $= 1\,\text{cm/s}^2 = 10^{-2}\,\text{m/s}^2$. The Earth's gravity field is approximately 980 gals, with variations about this figure usually given in milligals.

A simplified diagram of the gravity sensor is shown in Figures 4.4 and 4.5. The gravity sensor weight is shown as an approximately horizontal 'beam' which is free to oscillate about a horizontal axis. The weight end of the beam is supported by a diagonal spring whose upper end is adjusted by a micrometer screw to balance the pull of gravity on the beam. The beam is very highly damped to limit the beam motion caused by ship movement.

142 *High resolution site surveys*

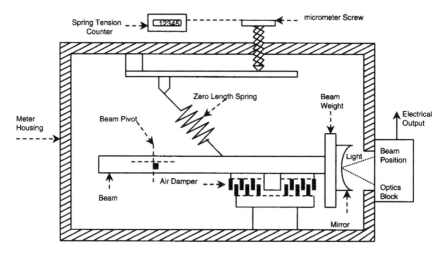

Figure 4.4 Simplified diagram of the Lacoste and Romberg gravity meter.

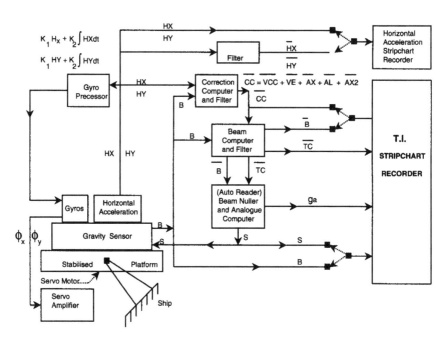

Figure 4.5 Schematic diagram of the Lacoste and Romberg gravity meter.

The platform is controlled by two high response torque motors, each motor being linked into two servo loops which are fed by outputs from a gyro and a horizontal accelerometer. The first loop is designed to null the gyro output, thus stabilising the platform in space. In the second, the accelerometer output

Non-seismic site survey techniques 143

is used to precess the corresponding gyro to maintain its verticality. The precession period must be longer than the ocean wave period and is usually chosen as four minutes. The system is specified to maintain a long-term platform verticality better than one second of arc through an angular range of ±30°. Any gravity measurement must take into account the vertical accelerations of the platform and this is done by averaging measurements, such that vertical accelerations are cancelled out. Vertical accelerations make it impossible to keep the beam stationary in its null position so it is very highly damped.

To measure gravity onboard a ship the acceleration due to gravity must be differentiated from accelerations due to the motion of the ship. Gravity is measured by first taking a static reading in port and setting the micrometer screw to a null point where the gravity beam is balanced. When the spring tension is correct a gravity reading can be taken. At sea the vessel movements, particularly vertical accelerations, make it impossible to keep the beam constantly nullified. The gravity sensor has, therefore, to be read with the beam in motion. The spring gravity sensor can read the position of the beam in motion. It has already been stated that the beam is highly damped so the accelerations can be ignored. The gravity sensor has a very high sensitivity so the position term can be ignored. The basic equation for reading gravity is given in Figure 4.6.

In Figure 4.6 the 'correction computer' uses the beam position B and the horizontal accelerometer outputs H_x and H_y to compute the relatively small corrections required because of possible small imperfections in the gravity meter. These individual cross-coupling corrections are added to form the total cross-coupling corrections CC. The time derivative of the average beam position, B', is added to this total cross-coupling correction CC to form the total correction TC. The total correction TC is fed into the automatic reader and combined with the spring tension S to compute a filtered gravity meter reading g_a. This filtered gravity g_a is recorded on the stripchart and magnetic tape recorders along with other variables used to monitor the total system.

Errors due to cross coupling between horizontal and vertical accelerations are inherent in the operation at sea and a correction is applied which is a function of beam position and horizontal acceleration. Another major error is the Eotvos effect, effectively the vertical component of the Coriolis acceleration. If a ship moves from west to east it has an angular velocity due to the ship's speed which adds to that of the Earth's rotation and increases the gravity reading. In the east to west direction the gravity reading is reduced. It is zero in the north–south direction.

Quality control checks

In terms of quality control, once the system has been set up the 'set beam zero and gain' check is performed. The gravity sensor is clamped and the beam position meter should be adjusted to zero. The second check is the 'manual

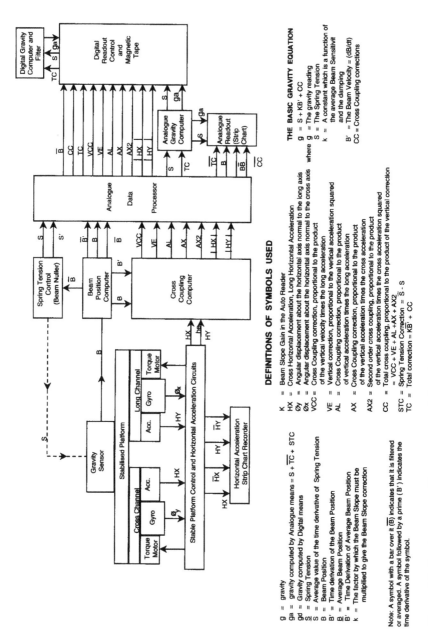

Figure 4.6 Gravity meter with computerised digital acquisition system.

base reading'. The beam position is set to zero and the average beam velocity is set to zero. The gravity servo should be operated and when the automatic reader comes to a steady reading it should be the same as that of the spring tension counter. The gravity and spring tension counters should then be checked against the corresponding stripchart records. The spring tension counter on the gravity sensor should then be checked with the spring tension counter on the automatic reader. They should be the same. Finally the calibration of the automatic reader should be checked. This calibration is usually referred to as the 'K-check' or 'average beam slope gain check'. The automatic reader is allowed to come to a stable reading which should coincide with the manual base reading. The spring tension counter should be slewed by 30 milligals and the counter should return to the base reading after 6–10 min.

4.9 Marine proton magnetometers

Magnetic sensing of the Earth's magnetic field as a means of prospecting for ore bodies has been used for over a hundred years. Fluxgate magnetometers were used for submarine detection in World War II. In the oil industry, magnetic sensing using proton magnetometers is a relatively inexpensive reconnaissance exploration method and was widely used in the 1960s and 1970s. In the 1980s the technique became rather less important but in the mid-1990s it again became a fairly frequent requirement on site surveys.

Marine proton magnetometers use a sensing element which is usually a bottle of lead-free kerosene around which is wound a coil of wire (see Figures 4.7, 4.8 and 4.9). Operation of the magnetometer[7] is based around the principle of nuclear precession. In the sensor unit, towed behind the survey

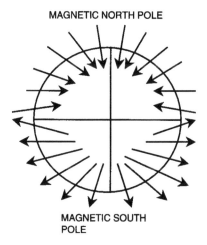

Figure 4.7 Marine proton magnetometer principles. Variation of inclination of the Earth's magnetic field.

146 High resolution site surveys

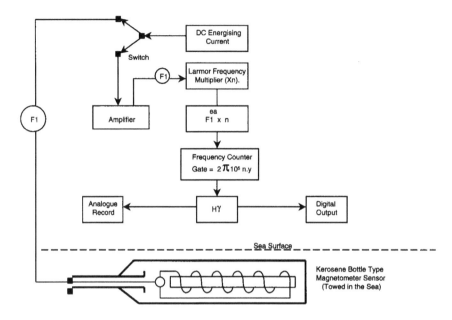

Figure 4.8 Schematic diagram of the Geometrics G801 marine survey magnetometer.

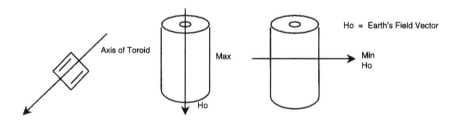

Figure 4.9 Sensor orientation in relation to the Earth's field vector.

vessel, protons exist in the bottle as hydrogen ions and these, due to the fact that they spin about a magnetic axis, have a tendency to be aligned parallel to the prevailing magnetic field. To make a magnetic measurement, a large current is passed through the coil circling the bottle to produce a strong magnetic field in a direction substantially different from the Earth's field, thereby aligning protons in this direction. The current is then cut off abruptly and the spinning protons revert into an alignment with the Earth's field. In doing so, these protons are subject to a precessional oscillation, the frequency of which is proportional only to the strength of the Earth's field. This relationship is described by Larmor's theorem:

$$H_o = 2\pi f / y$$

where H_o is the Earth's magnetic field in oersteds, f is the precessional or Larmor frequency and y is the gyromagnetic ratio 26,751.3. The precessional oscillation of the protons produces a small electrical signal in the coil of wire which can be detected and its frequency measured. Measurements cannot be made continuously but modern marine proton magnetometers can make measurements every four seconds or so to an accuracy of 0.5 gamma.[8]

Box 4.3
New units for old?
The units for expressing the Earth's magnetic field have caused some confusion over the years, with the various changes from fps/cgs to mks and then SI. The conventional unit of magnetic field intensity in the cgs/emu system was the oersted (Oe), which is effectively the same as the gauss (G), the unit of magnetic induction or flux density. The Earth's magnetic field is only about 0.5 Oe, so obviously the oersted is too large to measure the minute variations in the Earth's magnetic field from point to point on a site survey grid. For the measurement of terrestrial magnetism the gamma was used. The Earth's magnetic field is about 50,000 gammas or 50,000 nT (nanotesla). One gamma $= 10^{-5}$ G $= 10^{-5}$ Oe $= 10^{-9}$ Wb/m^2 $= 10^{-9}$ T. One gamma is, therefore, 1 nT and 1 nT is 10^{-5} G. The nanotesla is the modern unit of terrestial magnetism, replacing gammas. The weber (Wb) is the unit of magnetic flux in the mks system and is one joule/ampere. In the cgs system the unit of magnetic flux was the maxwell (Mx); 1 Wb $= 10^8$ Mx.

Typical magnetometers include the Geometrics G801 and G-866 marine proton magnetometers and Odom Seamag VIII marine proton magnetometer. Older systems include the Barringer magnetometer. Typical operating parameters might be as follows, while Figure 4.8 shows a simplified schematic of the Geometrics G801 survey magnetometer.

- Sensitivity: Variable, 1 gamma (nanotesla) at a 1.0 s sampling rate up to 0.125 gamma at a 10.0 s sampling rate
- Range: 20,000 to 100,000 gamma

Quality control checks

Quality control of the magnetometer includes ensuring that the in-sea sensor is towed well clear of the ship and that the ship's hull causes no magnetic anomaly that might mask seismic geological magnetic anomalies. As a rule of thumb the towing distance should be at least three times the survey vessel length. The sensor should be close to the seabed, about 3 m above the seabed to detect a buried telephone cable. Objects with finite length, such as wrecks, produce an anomaly that varies as the inverse cube of the distance from the

148 *High resolution site surveys*

sensor. Objects such as pipelines, with an infinite length, produce an anomaly that varies as the square of the distance from the sensor. The magnetometer repetition rate must be set correctly and a 3.0 or 6.0 s sampling rate is typical. It is preferable that the instrument be set up and checked for correct operation by running over a known source of anomaly such as a pipeline or wellhead. The magnetometer is often run as part of a site survey package and a very careful check should be made to ensure that no interference from other instruments is allowed to degrade the magnetometer record.

Generally speaking the background sea noise should be about 10 gamma close to the survey vessel. A 1000 ton ship should produce an anomaly of 0.3–0.7 gamma at 300 m. A 12 in. pipeline should produce an anomaly of 12–50 gamma at 24 m, a wellhead and casing 200–500 gamma at 15 m; a telephone cable 100 gamma at 3 m.[9]

Box 4.4
Gravity and magnetometer operators
Onboard a survey ship the gravity operator and the magnetometer operator are usually contracted personnel from small consulting groups or agencies. The stories told about these people are legion. One gravity operator did not know the difference between his gals (the unit of gravity) and his gammas (the unit of Earth magnetism), or his a*s* from his e*b*w. On another occasion a magnetometer just did not work at all. When towed over a known wellhead, seen on the echosounder record there was still no response. The operator refused to accept that his instrument was faulty. In an attempt to prove that it was working he applied even heavier impulses to the kerosene-filled bottle. 'Look', he said banging the bottle with a hammer, 'I am getting a response'. Of course it is easy to tell stories against people. The truth is that the boom-and-bust cycles in the oil industry ensure that nearly everyone with real experience is removed from the industry with every bust cycle. The lack of real experience across seismic field crews is a major problem.

4.10 Marine caesium vapour magnetometers

Caesium magnetometers have been in existence for a long time, but proton magnetometers were much more widely used in the 1970s and 1980s. In recent years a serious attempt has been made to develop a new generation of caesium instruments. A typical example is the Geometrics G-880 magnetometer. In common with its proton predecessor the caesium instrument provides a scalar measurement of the Earth's magnetic field. The measurement is expressed as the 'total field' intensity in nanoteslas or gammas and may range from 20,000 to 100,000 nT (0.2 to 1.0 G) at the Earth's surface. Local perturbations from geology or human-made objects add or subtract from the primary field as in a proton magnetometer.[10]

The caesium magnetometer operates on the principle that the energy of an atom is governed by the orbits in which its valence electrons revolve about its nucleus and also by the direction of the spin axes of the electrons themselves. In the case of caesium the caesium atom has only one electron in the outermost electron shell. Any electron has charge and spin and a consequent magnetic moment. The energy of the electron will therefore vary according to the direction of its spin axis relative to an ambient magnetic field vector. If the electron's magnetic field is aligned with the ambient magnetic field, the energy will be lower than if it was opposed to the field. Measurement of the energy changes caused by the electron's changing orientation allows a measurement of the Earth's magnetic field to be made.

Quantum physics states that an electron can take on only a limited number of orientations with respect to the ambient field vector. In the case of caesium there are nine. Each orientation will have a slightly different energy level. This is known as the Zeeman effect, which also states that energy differences from one Zeeman level to the next are roughly equal and always proportional to the strength of the ambient field. It is the energy differences between the Zeeman levels that are measured to determine the Earth's total field strength. The energy and frequency of a photon are related by Planck's constant and this fact allows a very accurate measurement of the energy differences by measuring the Larmor frequency, previously described.

The caesium vapour is placed in a chamber called an absorption cell. The cell is a 1 in. × 1 in. glass cylinder which holds the caesium vapour in a partial vacuum. There is a source of light called the 'lamp' (see Figure 4.10), which also contains caesium but at a higher vapour pressure. The light from the lamp does the actual pumping of the caesium atoms in the absorption cell.[11] On the far side of the absorption cell is a photocell for capturing the light that has passed through the absorption cell. It is this manner of feeding energy into the cell from a light source that is usually referred to as 'optical pumping'.

Polarised light is usually depicted as linearly polarised with the magnetic component horizontal and the vertical component electrical. Circularly polarised light differs in that the electric field runs like the thread of a screw. The circular polarisation can be either left or right handed. If a photon of incident light from the lamp has exactly the right amount of energy an electron can absorb it, moving the electron up to a higher orbit. With circularly polarised light this works better if the direction of spin matches the direction of polarisation. When circularly polarised light from the lamp source is shone through the absorption cell, the electrons that have a spin that matches the polarised light's direction will absorb the light and be kicked up to a higher orbit. When in this higher orbit the electrons are not stable and decay or fall back down releasing energy as light. Their spin direction becomes randomised in the process and the light given off when they fall is not aligned to the path of the absorbed light; for this reason the light passing through the cell will be dimmed slightly by the electrons absorbing it.

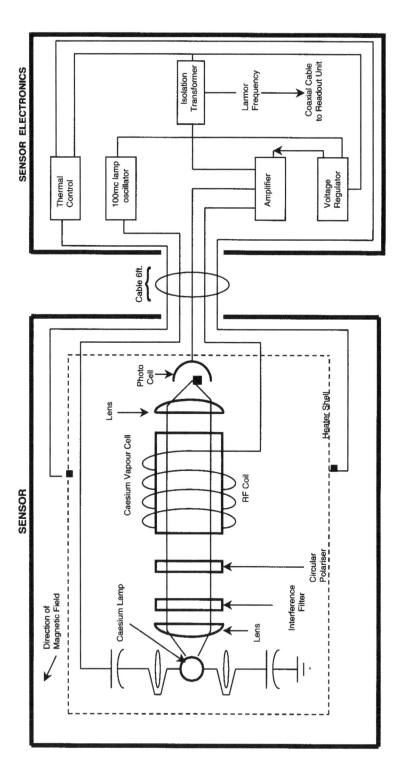

Figure 4.10 Marine caesium vapour magnetometer. (Reproduced from Milton and Dobrin, *Introduction to Geophysical Prospecting*, p. 515 by kind permission of McGraw-Hill Publishing Company.)

If the electron's spin axis is truly random when it falls back there is a chance that it will be aligned so that the light cannot kick it back to the higher orbit again. Over time all the electrons will eventually land with their spin axes in a manner that will not allow them to absorb light. When this happens the light passes through the absorption cell and impinges on the photocell. If radio frequency (RF) energy is beamed into the cell at just the right frequency to match the energy differences between orientations, the RF will tend to kick the electrons back over to other orientations, where they can again absorb light. The light will dim again at the photocell when the frequency of the RF 'depumping coil' is correct. This RF signal is called the H1 drive and the coil used to inject it into the absorption cell is the H1 coil.

Measuring the frequency of the RF signal produces an accurate reading of the ambient magnetic field. Sweeping the RF power back and forth allows calculation of the exact point where it couples with the electrons. This frequency can then be followed as it changes with variations in the magnetic field. This is the way a 'swept' vapour magnetometer operates. When electrons are kicked from one orientation to another they tend to do so in step with the RF signal. If a high frequency photocell is used, the light not only dims during the transition but acquires a slight RF modulation. If this RF signal is amplified it can be used as a depumping input into the absorption cell. By closing the loop in this way the whole system will oscillate at a frequency dependent on the ambient magnetic field strength. This is the principle of the self-oscillating alkali vapour optically pumped magnetometer.

An important operational consideration is the tolerance to movement of the caesium system compared to the proton system. The 'Doppler effect' from the rate of sensor rotation in rough seas is an order of magnitude less for caesium systems. A second important operational consideration is to ensure that the sensor is mounted so that the Earth's field angle is never less than 15° from the centre line of the sensor along both its length and width.

With reference to Figure 4.10, the light needed to move the electrons from one orbit to another must be exactly the right frequency; 3.3×10^{16} Hz is usually used and this corresponds to a wavelength of 894.35 nm. This light has a great deal of undesirable light which is removed by an interference filter. The lamp is about 0.2 in. in diameter and 0.4 in. long. The lamp is powered inductively because caesium is chemically reactive which means electrodes cannot be used inside the lamp. The induction coil operates at 80 MHz. The light rays from the lamp diverge so they are paralleled by the first lens. After filtering (already mentioned), the light passes to a circular polariser which is split along its diameter to make one side left-handed and the other right-handed polarisation. This prevents heading error caused by turning the sensor end-for-end which reverses the direction of travel of the light and has the same effect as reversing the polarisation direction. All sensors would have a 5 nT heading error if the polariser was not split along its diameter.

The cell is a glass chamber 1 in. in diameter and 1 in. long. It has a buffer gas and the caesium metal in it. The amount of caesium vapour is controlled by

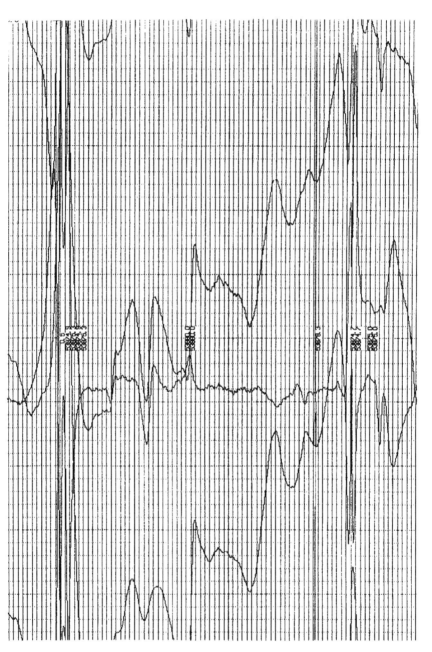

Figure 4.11 Geometrics marine proton magnetometer data. (Reproduced by kind permission of Geometrics (UK) Ltd.)

temperature means, typically at 55°C for the required amount of caesium vapour. A cell heater provides thermal control. A second lens then focuses the light onto the photocell, which is usually a wide area photocell with a good infra-red energy band response and a high frequency capability. The output signal is referred to as the Larmor frequency and will be about 0.1 mV at the photocell. This signal must then be amplified without introducing any phase shifts. Phase shifts cause the field measurement to be slightly offset.

Figure 4.11 shows marine proton magnetometer data.

References

1. UMEL Sargent gravity corer. Drawing and description taken from old contractor bid documents.
2. Penetrometer description taken from old site survey supervision reports.
3. Taken from old contractor bid documents.
4. Taken from old contractor data sheets.
5. This section is taken from old site survey supervision reports.
6. Jones, E.J.W. (1999), *Marine Geophysics*, John Wiley and Sons, Chichester, pp. 131–132.
7. Lacoste and Romberg, manual in the possession of the author, 20 pp. (It is unclear if this is a Lacoste and Romberg manual or a survey company manual.)
8. McQuillin and Ardus (1977), *Exploring the Geology of Shelf Seas*, Graham and Trotman, London, pp. 116–120.
9. UKOOA, *Conduct of Mobile Drilling Rig Site Surveys*, Volume 2, p. 66.
10. *Caesium Optically Pumped Magnetometers*, Geometrics Technical Report M-TR91, 8 pp. This description is almost completely that supplied by Geometrics.
11. Milton B. Dobrin (1976), *Introduction to Geophysical Prospecting*, McGraw-Hill Book Company, New York, p. 515.

5 Positioning systems

5.1 Introduction

As a historical backdrop to the problem of position fixing at sea it should be understood that prior to the introduction of differential mode global positioning, there were four categories of radio and electronic positioning systems.[1] The first of these was the hyperbolic method developed originally in World War II for aircraft navigation and ship positioning. The aircraft system was known as GEE[2] and the precursor system to the Decca Navigator was first used during the Normandy landings in June 1944. Various hyperbolic survey navigation systems were developed in the 1950s–70s. These included Raydist, Rana, Toran, Lambda and Decca systems such as Hi-Fix, Hi-Fix 6 and Hyperfix. Systems such as Pulse-8 and Loran were time comparison as opposed to phase comparison hyperbolic systems.

In hyperbolic systems the ship receiver is passive so that an unlimited number of ships can switch into the positioning chain. The survey vessel compares signals received from pairs of transmitting stations to derive a difference in distance of the survey vessel from these two transmitters.[3] Differences, usually expressed as integers in lane count between two adjacent pairs of stations, were plotted on a map as a family of hyperbolae for each station pair. A phasemeter enabled interpolation to be made between adjacent lines of zero phase difference but the problem always lay in identifying the hyperbolae themselves. The survey vessel had, therefore, to start from a known point and count lanes. It was very easy to lose the lane count and many surveys were degraded in consequence. Hyperbolic systems were usually operated at medium or low frequencies and suffered from radio wave effects such as sky waves at night and interference from other radio users.[4]

The disadvantages of hyperbolic systems ensured that range–range (circular) positioning systems were developed. An antenna onboard the ship transmitted a signal to a shore station which then retransmitted the signal to the ship, the round trip being accurately timed and converted into distance. The positioning chain had to be established for a particular survey and ranging systems tended to be expensive since only one or perhaps two ships could switch into the system. In their final form such systems could accommodate a number of ships by time-sharing of networks, with the actual number of ships dependent on position and fix update rate. Ranging systems became commonplace because of the increasing accuracy requirements for all types of survey work, particularly 3D surveys. Such systems included Shoran, Hiran, Maxiran, Trisponder, Syledis, Microfix, Argo, etc. Systems such as Argo and Syledis could be operated as either hyperbolic or ranging systems.

Acoustic systems represent the third category of system and are described in section 5.12 of this book.

The fourth category of system is the satellite positioning method. The development of satellite navigation systems started in 1959 when monitoring of the early Russian Sputniks revealed a very pronounced Doppler shift on the transmitted radio signals. The received frequencies were higher during the approach, since the source and observer were moving closer together and the frequency appeared lower as the source and observer moved further apart. It was realised that orbiting satellites whose position was known could be used as positioning beacons in space by measuring the Doppler shift in transmitted signals as the satellite moved across the sky. By 1964 the USA had the world's first satellite navigation system in operation. This system was known as Transit. The accuracy of an individual satellite fix was low (~ 30 m) and only about ten satellite fixes a day were possible. Satellite fixes were useful for calibrating hyperbolic lane count systems and for checking the results obtained from acoustic systems. Transit became open for commercial use in 1967.

Box 5.1
The ending of Decca Mainchain and the coming of RTK DGPS

At the time of going to press (May 2000) Decca Mainchain was finally switched off. This was the first commercial hyperbolic navigation system and was descended from the World War II system first used on D-Day in June 1944. The main advantage of the hyperbolic technique was that an unlimited number of ships could switch into the system. At the same time that Decca Mainchain was finally switched off, the successor company to Decca, Racal, announced that their new long-range real-time kinematic (LRTK) positioning system covered the entire North Sea to an accuracy of 20 cm in both position and height. In writing this book the author talked to numerous people in the oil industry about positioning at sea. No-one believed we would ever go back to terrestrial positioning systems. Satellite positioning systems are here to stay and represent the biggest leap forward in positioning at sea since John Harrison's chronometer back in the eighteenth century.

Transit has now been replaced by the American military satellite positioning system. There are a total of twenty-four satellites in the Navstar global positioning system. The satellites have two transmit frequencies, the L1 frequency at 1575.42 MHz (19 cm) and the L2 frequency at 1227.6 MHz (24 cm), modulated with two types of code and with the navigation message. The two types of code are the precision (P) code and the commercially available (CA) code. There are two types of observation, the first is the pseudo-range, which equals the distance between the satellite and the receiver plus a small corrective term for receiver clock errors. Given the geometric positions of the satellites, possibly best referred to as a precise mathematical description of the satellite orbit (satellite ephemeris), four pseudo-ranges are sufficient to solve, by simultaneous equations, positions in terms of x, y and z co-ordinates and to compute the receiver clock error. SV clocks are only healthy when stable and satellite vehicle (SV) clock bias is measured and included in the broadcast message. The second observation, the carrier phase, is the difference between the phase of the carrier signal transmitted by the satellite and the phase of the receiver at the time of measurement. Receivers are programmed to make phase observations at the same equally spaced times. In addition, receivers keep track of the number of complete cycles received since the beginning of measurement. This means that the actual output is the accumulated phase observable at pre-set times.[5]

The central problem with GPS used for seismic and site survey work is that the CA code is not accurate enough for survey work and the US Government is unwilling to allow unrestricted access to the P-code. At an early stage in the development of GPS it was realised that the P-code might not be available for commercial use and efforts were made to find GPS alternatives that upgraded the GPS CA code to the necessary survey positioning accuracy. By the mid-1980s it was realised that pseudo-range corrections, transmitted by a dedicated ground station, either terrestrially on MF/HF radio links or by commercial satellites, could overcome the CA problem and provide a differential GPS system with an accuracy of 3–5 m. The US authorities have now introduced selective availability (SA), to protect their military positioning capability. This has further degraded the accuracy of the CA channel. Fast update rates using commercially available communication satellites have largely circumvented this problem. A number of contractors now provide differential mode GPS. These include Racal Skyfix, DSNP (Dassault-Sercel Navigation Positioning) Veripos, and Fugro-Geoteam Starfix DGPS.

The Russian alternative to Navstar is the very similar GLONASS system (Global'naya Navigatsionnaya Sputnikova Sistema). There is a great deal of contemporary interest in combined GPS/GLONASS receivers in which a position fix is derived from use of both systems simultaneously. At present GLONASS offers fourteen satellites, with more planned. The disadvantage of GLONASS is that, with only fourteen satellites, observation of four or more satellites can be reduced to as little as 50 per cent of the time. The advantage of GLONASS is that it does not use selective availability and offers a more accurate stand-alone accuracy than GPS as a consequence.[6] General use of both GPS and GLONASS has been possible since the introduction of the Ashtech GG-24[7] receiver which combines GPS and GLONASS in one system.

Table 5.1 GPS and GLONASS comparison

	Position (m)	Velocity (knots)
GPS stand-alone	100	4.00
GLONASS stand-alone	20	0.05
GPS + GLONASS stand-alone	16	0.3
Differential GPS	0.90	0.1
Differential GLONASS	1	0.05
Differential GPS + GLONASS	0.75	0.1

This receiver effectively adds the GLONASS satellites to the GPS constellation by converting the GLONASS satellite positions to the GPS WGS-84 reference system. The receiver then treats all the range signals as if they were coming from an enlarged GPS constellation. GPS and GLONASS accuracy can be compared in terms of position and velocity (Table 5.1).[8]

With Navstar the GPS measurement capability is based on the measurement of carrier phases to about 1/100 of a cycle, which for real-time kinematic (RTK) survey equals 2–3 mm in linear distance. The two satellite frequencies penetrate the ionosphere relatively well. The time delay caused by the ionosphere is inversely proportional to the square of the frequency so carrier phase observations at both frequencies can be used to eliminate most of the ionospheric effect. The GPS satellites are in high orbits (26,000 km) and this ensures there is no atmospheric drag on the satellites. The only major error effect on the satellites' orbits are the Earth's gravitational field and the effects of the sun and moon which can be computed accurately. Solar radiation pressure on the satellite orbit and tropospheric delay on the signals are other forms of error and these can be compensated for, with the proviso that intense periods of sunspot activity can cause considerable problems.

5.2 Principles of GPS operation

The GPS satellites occupy six orbital planes, each having an inclination of 55° to the equatorial plane. There are three or four satellites in each orbital plane. The orbital path is close to circular with a semi-major axis of 26,000 km, giving an orbital period of slightly less than 12 h. The satellites complete two orbital revolutions while the Earth rotates 360° (one sidereal day) and the satellites rise about 4 min earlier each day. Because the orbital period is an exact multiple of the period of the Earth's rotation, the satellite trajectory on the Earth repeats itself daily.

The satellite transmissions are derived from a fundamental frequency of 10.23 MHz. Multiplying by 154 gives the first transmitted frequency of 1575.42 MHz. Multiplying by 120 gives 1227.60 MHz which is the second transmission frequency. The chipping rate of the precision (P) code is that of the fundamental code at 10.23 MHz. The chipping rate of the commercially

158 *High resolution site surveys*

available (CA) code is 1.023 MHz, which is one-tenth of the fundamental frequency. The P-code is a pseudo-random noise (PRN) code generated mathematically by mixing two other codes and does not repeat itself for 37 weeks. All satellites can, therefore, transmit on the same frequency and be distinguished because of the mutually exclusive code sequences being transmitted. The codes are updated once a week with the latest ephemerides, SV clock biases, etc. Since there are less than thirty-seven satellites some code blocks remain unused. These are available for transmissions from ground stations. It has already been stated that the CA code is only 1 ms long, the chipping rate being 1.023 Mbps (megabits per second). The timing sequences of the CA code are synchronised with the P-code and each satellite transmits mutually exclusive CA codes, thus making it possible to distinguish the signals received simultaneously from the satellites. The navigation message is modulated by the two carrier frequencies at a chipping rate of 50 bps. It contains information on the ephemerides of the satellites, GPS time, clock behaviour and system status messages. Relativistic effects are important in GPS surveying but can be accurately computed. The atomic frequency standards in the GPS satellites are affected by both special relativity (the satellite's velocity) and by general relativity (the difference in the gravitational potential at the satellite's position relative to the position at the Earth's surface).[9]

GPS time is set through the atomic master clock at the USAF Falcon Air Force Station near Colorado Springs, Colorado. GPS time is within 1 μs of UTC time, uncorrected for leap seconds.[10]

Each satellite transmits a navigation message. The full frame of the navigation message is 1500 bits long and is sub-divided into five subframes of 300 bits each. It takes 6 s to transmit one subframe at the chipping rate of 50 bps. The first subframe contains the coefficients to correct the onboard satellite time to GPS time. Subframes 2 and 3 contain the ephemeris for the transmitting satellite which provides sufficient data for co-ordinate fixing in WGS-84. Subframes 4 and 5 are each subcommutated twenty-five times so that a complete data message requires twenty-five full frames. The 25 versions of subframes 4 and 5 are referred to as pages 1–25 and repeat every 750 s. Subframe 4 is used for special messages such as ionospheric correction terms and coefficients for the conversion of GPS time to UTC time. Subframe 5 contains the ephemerides of up to twenty-four satellites. Page 25 of subframe 5 contains health data for all the satellites. The purpose of subframe 5 is to assist in the acquisition of signals from all other satellites. As soon as one satellite has achieved lock the navigation message can be read and the position of all the other satellites computed.[11]

Each satellite transmits and receives data to and from ground-based tracking stations. The satellites transmit the following data to the ground control stations:

- Uncorrected ephemerides, that is the orbital characteristics of the satellites;
- Uncorrected satellite time with respect to the GPS system time.

From the ground stations the satellites receive the following:

- Corrected ephemerides (orbital data);
- A satellite constellation almanac;
- Corrected satellite time.

A ship receives the following satellite data:

- Position fix data;
- Satellite 'health' data;
- Corrected ephemerides (orbital data);
- A satellite constellation almanac;
- Data for correcting ionospheric interference.

The satellite constellation almanac describes the times at which each satellite rises above the horizon relative to the ship receiving the message. The 'health' data is a quality control message and indicates the reliability of timing and fix data. The status of such data can be displayed and the operator decides whether to accept or reject the satellite data. To obtain an accurate position fix the shipborne receiver takes a time-of-arrival measurement on the satellite signal and uses the satellite ephemerides to calculate the position of the satellite being tracked at the exact moment of signal transmission. The time of arrival is determined by the clock bias of the signal and by synchronisation of the ship's receiver with the CA code generated by the satellite. The receiver then calculates a pseudo-range by scaling the sums of the signal propagation delays and clock bias by the speed of light. This pseudo-range is so called because it contains the clock errors of the receiver and satellite clocks. The true range is now calculated by taking into account the effect of clock bias on the satellite clock. Since at least four satellites are used, the clock bias for the user's clock can be found. The unknown quantities are x, y, z and t. Once clock bias is known, three-satellite navigation can give x and y positioning, while four satellites are required for x, y and z positioning. Velocity is calculated by measuring the Doppler shift on the carrier frequency. Each set of four Doppler measurements is accomplished by Kalman filtering.

Satellite GPS receivers onboard survey ships are fairly standard code correlation receivers able to generate within the receiver a replica of the codes with the chipping rate derived from the receiver's internal clock. Most civilian use (seismic survey, etc.) employs receivers that receive the CA code only. This is single frequency GPS operation on 1575.42 MHz. These receivers correlate the codes generated by them with the codes received from the satellites. If there is initially no match between the codes, the code generated within the receiver is shifted until matching is obtained. Once the signals are aligned, a code tracking loop, called a delay lock loop, ensures that both code sequences remain aligned. The time shift in the two sequences of codes is a measure of the travel time of the signals from the satellite to the receiver.

Multiplying this time delay by the speed of light gives the pseudo-range (already described). This range measurement is contaminated by clock errors in the receiver as well as clock errors in the satellite, hence the name 'pseudo'. Once lock has occurred the receiver can read the navigation message and make use of GPS time and the ephemerides. In addition to providing pseudo-ranges, the receiver also provides the carrier phase observable, that is, the reconstructed carrier phase. This is the difference between the phase of the incoming Doppler-shifted carrier wave from the satellite and the phase of a reference signal generated by the receiver at pre-set times. The carrier phases can be measured to an accuracy of 2 mm or so. The reconstructed carrier phase can be related to the satellite–receiver distance. A change in the satellite–receiver distance of one wavelength of the GPS carrier results in a one-cycle change in the reconstructed carrier. The reconstructed carrier phase changes according to the continuously integrated Doppler shift of the incoming signal.

In sum, a GPS receiver should measure the pseudo-range (code phase) measurement, the instantaneous Doppler, the time tag information for the measurements, the standard deviation of the pseudo-range measurement, the standard deviation of the instantaneous Doppler, the signal-to-noise ratio, the almanac and epheremis data, the alpha and beta parameters (discussed later), and the conversion between GPS and UTC time.

5.3 Co-ordinate systems

A local survey is usually performed with reference to previously established geodetic control stations, which in turn are defined with respect to the local datum. A satellite-based positioning system operates on a global basis and cannot, therefore, provide fix results in each local datum. In consequence a local datum is adopted and position fixes must be transformed from the global datum to the appropriate local datum. Oil companies are usually granted exploration licences based on block areas defined in a local datum, specific to the country that issues the licences. It is up to the oil company to know the co-ordinate transformation from the local datum to the satellite datum, though a designated survey company may establish this on behalf of the oil company.

Any point on the Earth's surface can be defined by the co-ordinates x, y and z. The longitude of any point is the angle between the x-axis and the point (x, y) projected onto the equatorial plane. The latitude of the point (x, y, z) is more complex, being geodetic latitude rather than geocentric latitude. The geodetic latitude is the angle between the equatorial plane and a line through (x, y, z) which is normal to the spheroid at that point. This definition of latitude is influenced by the shape of the spheroid. Altitude is normally surveyed with respect to mean sea level and the geoid is the surface that describes mean sea level over the entire Earth.[12] The geoid is a gravity equipotential surface that coincides with mean sea level. A spheroid is derived to be the best fit in some way to the true. The smooth spheroid and the rather lumpy geoid deviate from each other, the difference being known as geoidal height. A spheroid is an ellipse rotated about its minor axis. The spheroid represents the nearest simple mathematical shape to the geoid.

It would be convenient if all the world used the same spheroid for mapping. Unfortunately, many spheroids are in use, each one being designed to be a best fit to the true geoid in that particular area. Any datum is based on many surveyed points, surveyed with respect to one another and positioned independently by astronomical observations. All data are combined to establish a best fit datum having an origin point represented by a physical marker for which there are defined values of latitude, longitude and height.

If two spheroids associated with different datums were drawn, not only would the semi-major axes and flattening coefficients be different but their centres would be at different positions. In consequence, the co-ordinates of a position are dependent on which datum is being used and the datum should always be specified when listing co-ordinates. By means of satellite measurements (satellite geodesy), the offset of one datum with respect to all others can be compared. The world-wide datum now in use is WGS-84, established for the Navstar GPS system. By contrast the GLONASS system uses the PZ 90 datum.

When comparing local survey data with satellite fix results, the effect of datum shift must be considered, but when there is a difference it may not be totally due to datum shift. Too often, after a difference is found, the local survey is checked and found to be in error. A local system may show excellent precision and consistency of results and still be quite wrong because of survey error. If both a local system and a satellite system are used, a calibration of position difference before the survey begins is extremely wise. The offset is a measure both of datum shift and of local survey error and can be applied as a basis throughout the survey. Having discussed geodesy in general terms, some basic surveying definitions can be arrived at.

5.4 Geoids, ellipsoids, datums and shifts

The geoid is the gravity equipotential surface which coincides with the mean sea level. This surface is fairly even and close to being ellipsoidal in shape. The maximum departure from a best-fitting ellipsoid is approximately 110 m. The reason for this difference is the effect of the Earth's rotation and the fact that gravity is varying from location to location depending on irregularities in the Earth itself. The geoid is fundamental in geodesy because so many observations are related to it. Observations made on the Earth's surface are reduced to sea level, that is, to our geoid.[13]

The ellipsoid can be represented as a surface (spheroid) which is an ellipse rotated about its minor axis. The ellipsoid represents the nearest simple mathematical shape to the geoid. It is important to understand that the ellipsoid is a mathematical model and does not, unlike the geoid, physically exist. The shape of an ellipsoid is normally defined by its semi-major axis and

162 *High resolution site surveys*

Box 5.2
How to lose an oilfield in one easy lesson

Once upon a time there was an oil company who had a nice little oilfield somewhere off the coast of Africa. There was a civil war in the country off whose coast the oilfield was situated and the oil company had to withdraw their personnel and rigs. Some years later, as things settled down again, the oil company went back into the oilfield and could not find it. How can we lose an oilfield? raged the senior vice-president and the rest of the board. How indeed, gentle reader? This is what happened. The oilfield was surveyed and the rigs were positioned using a particular radio-positioning system with a particular datum and spheroid. When the oil company recommissioned the oilfield they decided that the old co-ordinate system was inaccurate and full of mistakes, which as it happened was absolutely true. They decided to use a newer and much more modern radio positioning system and bring the field position into line with satellite navigation. The co-ordinate system in use would be thoroughly modern, with no mistakes or errors. The much more modern radio-positioning system was installed with new reference station positions, etc. Of course the 'lost' oilfield was not really lost at all and was 'rediscovered' by going back to the old position reference co-ordinates, mistakes and all. Checking continuity in co-ordinate systems from the seismic and site survey work to oilfield commissioning sometimes brings surprising results. It is very important to repeat mistakes if they have been made. Accuracy is one consideration in positioning but repeatability, the ability to return perhaps years later to a particular point on the Earth's surface, is just as important. Repeatability means repeating mistakes. Satellite navigation has superseded all the old terrestial positioning systems but GPS positioning results in WGS 84 still have to be converted to the local co-ordinate system. The potential for mistakes is still as great as ever.

---

inverse flattening. Figure 5.1 shows the various classical geometrical formulae that have come the way of the author over the last 25 years. If more information is required, 'Bomford's Geodesy' is recommended.[14]

The terms spheroid and datum are sometimes confused. The name of the spheroid will tell you which model of the Earth is being used, normally defined by the semi-major axis and the flattening. To establish a datum involves positioning and orientating this spheroid inside the geoid in such a way that the surface of the spheroid (the ellipsoid) as closely as possible matches the surface of the geoid in a particular part of the world. Each local datum is also identified by its origin, a point at which latitude and longitude are defined. The American datum uses the Clarke 1866 spheroid located by its

REFERENCE SURFACES
  a. Solid Earth
  b. Equipotential surface
  c. Geoid
  d. Plane, Sphere, Spheroid (Ellipsoid of Rotation)
  e. Plumb line, Vertical, Normal, deviation of the Vertical
  f. 'First' and 'second' Geodetic problems

GEOMETRY OF THE ELLIPSOID
  a. Definition of the Ellipsoid

$$\frac{x^2}{a^2} + \frac{y^2}{a^2} + \frac{z^2}{a^2b^2} = 1$$

SEMI-MAJOR AND SEMI-MINOR AXES a and b

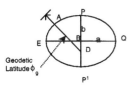

Geodetic Latitude ϕ_g

FLATTENING, 1st and 2nd ECCENTRICITY

$$f = \frac{a - b}{a} \quad \text{Therefore } b = a(1 - f)$$

$$e^2 = \frac{a^2 - b^2}{a^2} = 2f - f^2$$

$$e'^2 = \frac{a^2 - b^2}{b^2} = \frac{e^2}{1 - e^2}$$

RADII OF CURVATURE

$$\text{Meridional } \rho = \frac{a(1 - e^2)}{(1 - e^2 \sin^2 \phi)^{3/2}}$$

$$\text{Prime Vertical } \gamma = \frac{a}{(1 - e^2 \sin^2 \phi)^{1/2}} = AD$$

$$\text{Any Other } R_\alpha = \frac{\rho \gamma}{\rho \sin^2 \alpha + \gamma \cos^2 \alpha}$$

ASTRONOMIC (TRUE) AND GEODETIC (SPHEROIDAL) CO-ORDINATES

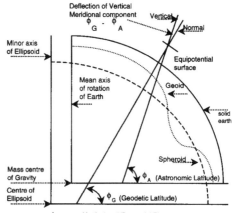

ϕ_A = Vertical and Equatorial Plane
ϕ_G = Normal and Equatorial Plane
True Meridian = Vertical and major axis of rotation
Geodetic Meridian = Normal and minor axis of ellipse
Astronomic Longitude λ_A and Geodetic Longitude λ_G
λ_A = True meridional plane, true Greenwich meridian
λ_G = Geodetic meridional plane, true Greenwich meridian
Astronomic azimuth (α_A), Geodetic azimuth (α_G)
α_A = True meridional plane and second point.
α_G = Geoidal meridional plane and second point

DEFINITION OF A 7 - PARAMETER GEODETIC SURVEY ORIGIN
  (a) Size of Spheroid (a and e^2)
  (b) Attitude of the Spheroidal minor axis (parallel to the Earth's axis)
  (c) Observe $\phi^{(0)}_A$, $\lambda^{(0)}_A$ and $h^{(0)}$ (Height above Geoid)
  (d) Define $\phi^{(0)}_G$, $\lambda^{(0)}_G$ and $N^{(0)}$ (Geoid - Spheroid separation)
  (e) Normally $\phi^{(0)}_G = \phi^{(0)}_A \quad (\xi = 0)$
  $\lambda^{(0)}_G = \lambda^{(0)}_A \quad (\eta = 0)$
  N^0 = Zero $\{\text{Therefore } (N + h^0)\}$

Figure 5.1 Classical geometrical geodesy – definitions and formulae.

Table 5.2 Well known ellipsoids

Name	Year	a (m)	f	Usage
Everest	(1830)	6,377,276	1/300	India
Airy	(1830)	6,376,542	1/299	Great Britain
Bessel	(1841)	6,377,397	1/299	Japan
Clarke	(1866)	6,378,206	1/295	North America
Clarke	(1880)	6,378,249	1/293	France, Africa
International	(1924)	6,378,388	1/297	Europe
IAG Lucerne	(1967)	6,378,245	1/298	Russia
Krassowsky	(1940)	6,378,160	1/298.247	IAG recommended
WGS '72	(1972)	6,378,135	1/298.260	Transit satellite navigation
WGS '84	(1984)	6,378,137	1/298.2572	GPS satellite navigation

point of origin at Meades Ranch, Kansas. The International 1924 spheroid is located in Potsdam and constitutes the European datum of 1950 (ED50). It should be understood that the same spheroid can be used in different datums by locating the centre of the spheroid at different positions. For example, the International spheroid of 1924 is used both by the ED50 datum and the Argentine datum but each has obviously a different origin point. Another example is the Australian and other South American datums which use the IAG Lucerne 1967 reference ellipsoid but, again, with different origins.

Establishing a datum has the effect of determining the centre of the spheroid. This means shifting the centre a certain number of metres (delta x, delta y and delta z) from an established reference, nowadays normally the centre of a spheroid associated with one of the satellite datums. Using a simple Cartesian shift is usually referred to as a three-parameter shift. This assumes that the three axes of the spheroids are parallel. The term Cartesian means using a rectangular co-ordinate system with a common origin. A five-parameter shift will be more accurate. The two additional parameters are the rotation about the z-axis which is parallel to the nominal rotation axis of the Earth and the application of a scale factor to the length of the vectors. An even more accurate seven-parameter shift is available with the two additional parameters being rotations about the x- and y-axes. It should be understood that spheroids can only be compared in terms of seven-parameter datum shifts.

Box 5.3
Three-parameter, five-parameter and seven-parameter shifts

For a long time most site survey operations used simple three-parameter shifts as did 2D and even 3D seismic surveys. Then one contractor introduced an integrated navigation system with a seven-parameter shift. 'Absolutely essential for 3D seismic surveys,' said the contractor, who was absolutely right. For very limited site survey work a three-parameter shift is acceptable; where the numbers of parameters becomes significant is when the survey lines start to lengthen. This author has seen a big 3D seismic survey where the lines were over 25 km in length. The integrated navigation system could handle three parameters only. 'It does not matter', said the contractor; 'we agree' said the oil company. In fact, significant distortions do take place on long lines if three parameters only are available. It is some years since this type of operation took place and these days, virtually every seismic and site survey operation will use a seven-parameter shift.

A satellite datum can cover the entire planet but still requires a spheroid and a point of origin. By observing the influence of the Earth's gravity field on the orbits of satellites it was possible to develop a model of the geoid. The first manned spaceflights in Project Mercury required tracking stations and orbital maps in a uniform system. The result was the Mercury datum (1960). Further developments of geocentric datums led to WGS-72 and WGS-84, and both are geopotential models.[15] Each successive model refined the size and shape of the spheroid to best fit the entire Earth. The centre of the spheroid is at the Earth's centre of mass.

Any position is usually referenced to the surface of the spheroid but there is also an altitude component involved because the geoid does not exactly match the spheroid. The deviation between the two surfaces is known as the geoidal height. There are also undulation models such as O5491A.

5.5 GPS differential mode principles

When GPS was first conceived it was decided that the P-code would be encrypted and available only to military and other privileged users, while the CA code would be free of encryption for general commercial operations.[16] It was also intended that CA code users would be required to pay a licence fee for use of the CA code. In the event the achievable accuracy of the CA code proved to be much greater than expected. In consequence the US Government introduced selective availability (SA), to protect their military interests. The ephemeris broadcast on the CA code is deliberately degraded so that the satellite positions are less accurate. The clock frequencies are jittered so that the clock bias between the satellite and the receiver is continually changing. These problems can be overcome with differential mode of operation, even though in a stand-alone mode GPS is degraded to an accuracy of ~ 100 m, perhaps more when SA is switched on. Only the newer Block II satellites are fitted for SA. The Block I satellites were all launched between 1978 and 1985, with the Block II satellites launched in 1989–1990. A second series of Block II satellites, designated Block IIA, were all launched in the period 1990–1999. The latest generation of satellites, Block IIR, will keep GPS going far into the twenty-first century. At the millennium there were no Block I satellites remaining in service.

The fundamental principle of DGPS is the comparison of the position of a fixed point, referred to as the reference station, with positions obtained from a GPS receiver at that point. The observed difference can either be considered as a two- or three-dimensional geographical co-ordinate offset or as a series of corrections to the satellite range data. Position differential corrections are not much use since the survey vessel must use the same satellites as the reference station. The pseudo-range technique computes a correction to the range of each satellite observed at the reference station but does not need all of the same satellites to be observed at the reference and mobile stations, provided there are enough common satellites for a fix.

Medium frequency (MF) radio beacons were first used for differential corrections, operating at 285–325 kHz. A message format known as RTCM SC104 (Radio Technical Commission for Maritime Services Special Committee 104) has become the industry standard for encoding DGPS corrections. There were severe limits imposed on MF and later high frequency (HF) corrections.

MF/HF links were prone to sky-wave interference and were very weather dependent. Finally, there was no way of guaranteeing that the satellites observed at the reference station were the same as those observed at the mobile station. This can be a fatal error because the position correction passed to the survey vessel is completely dependent on which satellites are being tracked, though it should be noted that this statement relates to an x, y, z position correction and not to RTCM 104 pseudo-range corrections. It should also be noted that MF corrections have now become obsolete with selective availability. Most DGPS systems now receive their differential corrections by satellite linkage, using communication satellites for this purpose. Fast updates at typically once a second are used, as opposed to updates at once every 30 s on MF correction systems. A typical modern onboard differential installation is shown in Figure 5.2.

Using pseudo-range corrections the reference station computes the difference between the measured and the theoretical pseudo-range. This difference is the pseudo-range correction and is transmitted to the mobile user as shown in Figures 5.2 and 5.3. The mobile user then corrects his range measurements by the correction received from the reference station. The advantages of a pseudo-range correction is that the reference station transmits corrections on all satellites above the horizon. The mobile station is, therefore, left with a large degree of freedom to choose between all available satellites.

A DGPS reference station will include satellite receivers to track the GPS satellites. It will also include a real-time computer which reads pseudo-range and ephemeris from the GPS receivers and calculates a pseudo-range correction which is transmitted to the survey vessel. Onboard the survey vessel the computer reads the pseudo-range measurements and ephemeris from the GPS receiver. It takes correction data from the reference station and computes a differentially based GPS position. This data is stored on disk or mini-floppy for post-processing. The final position solution is based on the filtered pseudo-range and the degree of filtering is usually operator selectable. The differential correction is updated every few seconds and again the time between updates is usually operator selectable. In the time between corrections the last differential update is used and the onboard software predicts the corrections needed up to the next update.

5.6 Real-time kinematic (RTK) GPS

The major GPS innovation taking place today concerns a method of carrier-phase differential GPS positioning that gives centimetre accuracy.[17] This technique is often referred to as RTK GPS. The carrier phase is the phase of the received carrier with respect to the phase of a carrier generated by an oscillator in the GPS receiver. That carrier has a nominally constant frequency,

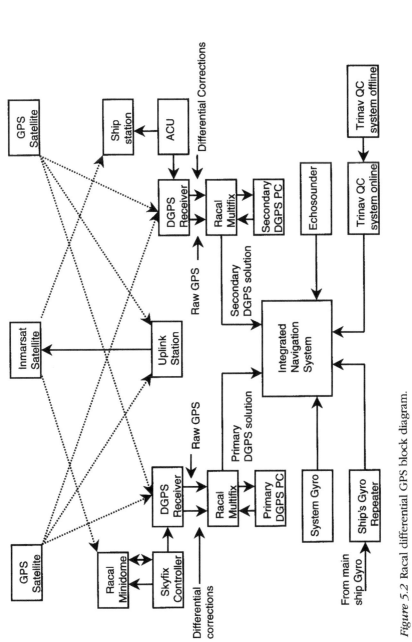

Figure 5.2 Racal differential GPS block diagram.

168 *High resolution site surveys*

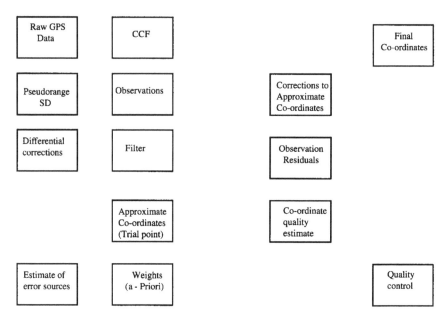

Figure 5.3 Multifix DGPS program overview.

whereas the received carrier is changing in frequency because of the Doppler shift induced by the relative motion of the satellite and the receiver. The received carrier's phase is related to the phase of the carrier at the satellite through the time interval required for the signal to propagate from the satellite to the receiver. The observable carrier phase will be the total number of full carrier cycles and fractional cycles between the antennae of a satellite and a receiver at any instant.

A GPS receiver has no way of distinguishing one cycle of a carrier from another. The best it can do is to measure the fractional phase and then keep track of phase changes. The initial phase is undetermined by an integer number of cycles. To use the carrier phase as an observable for positioning, this unknown number of cycles, or ambiguity, must be estimated along with other unknowns including the receiver's co-ordinates. If we convert the measured carrier phase in cycles to equivalent distance units by multiplying by the carrier's wavelength, the carrier phase observation can be expressed in a form similar to a pseudo-range. The carrier phase will have some ambiguity present but can be treated as a pseudo-range. All GPS receivers must lock onto and track the signal's carrier to measure pseudo-ranges but they do not record carrier phase observations for external use. Some can use carrier phase measurements to smooth the pseudo-range measurements. The carrier phase rate of change is related to the Doppler shift, which is used to determine velocity.

For a high accuracy positioning, carrier phase measurements made by one receiver are typically combined with those made by another receiver to form double differences in which the effects of satellite and receiver clock errors are eliminated. These double differences are then processed using a least squares filter to estimate the relative co-ordinates of one receiver with respect to the other. If the co-ordinates of one receiver are known, because it is at a known geodetic control point, then the co-ordinates of a second receiver, onboard a survey vessel, can be calculated in the same frame. In the early 1980s carrier phase measurements were first used for precise static positioning but this took many hours. In modern receivers the RTK technique can be refined to continuous position fixing using carrier phase techniques.

Swath bathymetry often uses RTK positioning, as does ROV survey. Tidal corrections can be directly derived from RTK data, without the need for tide tables. A disadvantage of RTK is that a very wide band data link is required.

5.7 DGPS errors

The choice of two radio frequencies is governed chiefly by ionospheric and atmospheric errors which are caused when the signal path from the satellite to the receiver passes through the ionosphere and troposphere. The ionosphere is a layer of ionised air some 50–500 km above the Earth's surface.[18] The troposphere is the lower part of the atmosphere and extends from ground level to 9–16 km above the Earth. In the troposphere changes of air pressure, temperature, density and humidity are all important. Calculation of the relative speeds of two different signals minimises the ionospheric and tropospheric effects. Ionospheric effects are usually considered to cause an error of ~ 5 m, while tropospheric errors produce errors of ~ 1 m or so. Ionospheric errors of greater than 5 m, up to 10–15 m, can be expected as the sunspot cycle reaches a maximum in 2001. This remark applies particularly to surveys conducted in equatorial regions.

Multipath errors, caused by reflected signals from near the satellite receiver are often detected. The satellite signal, instead of following a straight path to the receiver's antenna, is bounced off various local obstructions. This increases the effective length of signal received from the satellite to the receiver. These multipath signals confuse the receiver's calculations and cause wave interference with the direct signal, producing a confused and noisy result. This type of interference is sometimes referred to as 'ghosting'.

Another source of error is satellite clock bias error, caused by minute discrepancies in the accuracy of the atomic clocks used in the satellites. Satellite clock errors can degrade the final calculated position by ~ 1.5 m. DGPS should eliminate clock errors.

The exact position of every orbiting satellite within the positioning constellation is constantly monitored but very small positional or ephemeris errors will always occur. These errors can occur between specific monitoring periods and produce errors of ~ 2.5 m. Generally, the wider the spacing between the satellites being tracked by a GPS receiver, the smaller the position error. As there are often more satellites available than are actually required, some receivers select the few nearest satellites and ignore the rest. Geometric dilution of position (GDOP) is present when the receiver selects satellites that are close together and the intersecting positions at the receiver are very similar, crossing at shallow angles. Again, DGPS should virtually eliminate ephemeris errors.

In considering errors, two different kinds of error in a derived DGPS position can be said to exist. These are random errors and biases.[19]

Random errors are by definition unpredictable. Atmospheric scintillation (which causes stars to twinkle) and electronic noise causing imperfect code cross-correlation are considered to be random errors. As the 11-year sunspot cycle reaches a peak of activity in April 2000 the possibility exists that scintillation errors will increase to a point where at dawn and dusk in tropical regions DGPS error may rise to 5–15 m. Such errors are covered by the general heading of 'stochastic quantities'. In terms of mathematical analysis such errors are considered to follow a normal (Gaussian) distribution.

Biases are errors not covered by the statistics used to describe random errors and take two forms, gross errors and systematic errors. Systematic errors can be completely predicted and are removed or accounted for by careful calibration and modelling. Multipathing at a reference site is a good example of systematic bias and once completely known it can be predicted from the satellite constellation. Multipath errors change with the satellite position. Gross errors are likely to occur when, for example, a new, very low elevation satellite is included in the position fixing equation, or the incorrect height is entered into the navigation computer.

When DGPS is used the term precision is used to describe the quality of a DGPS fix with respect to random errors. A very precise fix is one where random errors are known to be small. Such a fix is said to be of high precision. The term reliability is used to describe the quality of a DGPS fix with respect to bias errors of the type described above. A highly reliable fix is one in which quite small biases are detected, whereas in an unreliable fix, large biases are not detected. It should be pointed out that precision is not exactly the same as accuracy.

In order to convert differentially corrected pseudo-ranges into positions two mathematical models are used, a functional model and a stochastic model. The functional model describes the mathematical relationship between the measurements and the required position. The stochastic model describes the statistical quality, that is, the precision of the measurements. The functional models used for DGPS are virtually the same for all receivers in all situations and incorrect functional models cause biases of the systematic error type. Stochastic modelling is extremely complicated because it involves a differentially corrected phase-aided pseudo-range. This in turn is determined by the satellite elevation angle, the satellite acquisition time, the state of the atmosphere, the distance to the reference station, the latency of the differential corrections, the receiver noise characteristics and the multipathing environments at both the reference and mobile stations.

Most DGPS system providers have spent a great deal of time and money developing their own proprietary algorithms and do not provide exact details of these algorithms. Once the functional and stochastic models are known, fixes are usually computed by a process known as least squares. Precision is usually assessed by means of a standard deviation, sometimes referred to as variance, though the two are not exactly the same. The standard deviation is a measure of the spread of any random errors remaining in any component of the position. A population of errors is normally described as a probability density function (PDF). This is a mathematical function which, when integrated, gives the probability of an individual error falling between specified bounds.

When considering horizontal positional standard errors for offshore surveying it is not sufficient to look at errors in the two main orthogonal directions of east–west and north–south. The errors in all directions must be considered, usually known as the error ellipse, which is an approximate graphical representation of the standard deviations in all directions. The semi-major axis of this ellipse lies in the direction of lowest precision (highest standard deviation) and, conversely, the minor axis shows the direction in which the fix is strongest.

Reliability is measured by means of marginally detectable errors (MDEs). These are bias errors sometimes referred to as 'outliers'. A test statistic for bias error might be arbitrarily set such that in 1 per cent of cases, it is assumed that the observation was generated by an incorrect process. This statistic is the level of significance and denoted by the Greek letter alpha. In theory, alpha per cent of good data is always rejected to ensure that any bad data is also rejected in the process. When bad data is accepted a type two error is said to have occurred, known as beta error.

The internal reliability of the fix is measured by an MDE as follows. If values of alpha and beta were chosen at 1 per cent and 20 per cent, then an MDE of, say, 2.7 m can be calculated, though this can vary from fix to fix. From this it can be stated that when a bias error is carried out with a level of significance of 1 per cent then there is a 20 per cent chance of a bias error of 2.7 m remaining undetected. External reliability is assessed by the largest horizontal positional MDE. A positional MDE is simply the effect of a marginally undetected bias on position. If an MDE occurred at a 20 per cent level of detection, then an error in position of 13–14 m could occur. Within least squares all statistical testing is based on residuals. These values are added to measurements to make them geometrically consistent with the least squares estimate of position, that is they are corrections to measurements. Fairly obviously, consistently large MDEs should be avoided.

The w-test is used to identify biases in the data. If the unit variance indicates that all models are correct, the measurement concerned is rejected and the least squares estimation of position is repeated. The w-statistic is obtained by

172   *High resolution site surveys*

dividing a residual by its standard deviation, often referred to as the normalised residual. The unit variance statistic is derived from the weighted sum of the squares of the residuals and the normally accepted figure is unity.

For quality control it is essential to assess the precision and reliability of each position fix. The w-test and F-test (unit variance test) should be carried out for each fix. The precision (a posteriori error ellipse) and reliability (positional MDE using a power test of 80 per cent) should also be calculated. Again, it should be emphasised that UKOOA guidelines for DGPS are the industry standard for quality control and are adhered to in the software routines used by virtually all contractors.

5.8   Skyfix, Veripos and Starfix

At the time of writing (summer 1999) there are three leaders in the field of DGPS operations, Racal Skyfix, Dassault-Sercel Veripos and Fugro-Geoteam Starfix. The descriptions given here do not attempt to favour one system against the others, merely to describe the three main systems in use. A typical onboard positioning system will include two GPS receivers and a differential corrections receiver interfaced to the Inmarsat link. The GPS signals and differential corrections are then integrated to produce a position solution. The position solution is then used for vessel positioning and steering along the pre-plotted survey lines. It is sometimes the case that a streamer tailbuoy tracking system will be used. These units are usually referred to as RGPS (relative global positioning system).

Skyfix

The Racal Skyfix system represented a major advance when first commissioned because the differential data is transmitted by satellite on Inmarsat satellites. After 6 years in operation there are approximately 1000 users a day. A network of about seventy world-wide stations are linked by dual land-lines to control stations, with permanent data communication links to the Aberdeen control centre. The Skyfix reference stations are established on the ITRF91 based WGS-84 co-ordinate system. Each reference station is equipped with dual GPS receivers, PCs and software. The Skyfix Mark 3 demodulator/decoder unit uses an integrated Trimble DSM Eurocard GPS receiver and a PC/AT compatible 386/486 processor card, running a virtual base station (VBS) program. The system software produces a composite RTCM stream based on a selection of up to six pre-selected reference stations, for direct injection into the DSM receiver or output to an external GPS receiver. The wide area network computation is performed by an internal PC card running VBS software, developed by Racal Survey. The VBS software provides a multiple reference station GPS solution for direct injection into the internal GPS receiver in a Skyfix 90938 unit. The VBS program generates RTCM corrections for the user location, based upon information from a multi-reference station stream and a GPS receiver. The receiver is used to supply GPS positioning message information and an approximate user position.[20]

Data from up to six reference stations creates a wide area network. The data from each reference station is assessed for quality and integrity. The valid data is then combined, forming a clock model and deriving the component errors of the GPS. These are the fast clock and selective availability corrections, also slow ionospheric and orbital components. Having derived the various corrections, integrity checks are carried out to validate the data and a mapping function creates the parameters to enable the system to be used over a wide area.

The differential messages are uploaded at the control stations on the dedicated Inmarsat channels to the various Inmarsat satellites. Onboard the survey vessel an Inmarsat receiver decodes the incoming correction signals. The Skyfix data rate at 1200 bps enables corrections to be sent simultaneously for a number of reference stations at time intervals of 3–4 s. This is fast enough to overcome the current level of selective availability. The datalink operates in the L-band at 1.5 GHz and the availability of the datalink is better than 99.998 per cent of the year. Skyfix integrates successfully with most integrated navigation systems.

Veripos

Sercel produced their own DGPS system, known as Aquapos, which used MF radio wave transmissions for the differential corrections. Sercel are now a subsidiary of Dassault-Sercel Navigation Positioning (DSNP) and have produced a successor system known as Veripos, with Aquarius as their long range kinematic system. Veripos was established in 1990 as part of a consortium which includes Subsea Offshore, Ormston Technology and Osiris B.V. Veripos is offered in two variants, one using Inmarsat differential corrections, Veripos-I, and Veripos-HF which uses high frequency (HF) differential corrections. As with any DGPS system the critical factor is the latency of the differential data and the reliability of the data link. High latency means high correction age and poor position instabilities.

One of the advantages of an HF differential link is that the corrections are transmitted directly from a reference station and this ensures that latency is minimal, less than 1 s, with correction update rates at less than 4 s and a data age below 5 s. HF radio transmissions suffer traditionally from fading and this is minimised by using two HF transmission frequencies in the 1.5–3.5 MHz band. Different correction data is modulated onto each frequency and the transmissions are displaced in time so that when both are received in parallel, the correction update rates are halved. The DSNP NR203 receiver uses a 50 baud data rate that gives coverage up to 700 km. The Veripos HF differential corrections used to be transmitted in a proprietary format but in 1997 Veripos R was introduced to provide an RTCM SC104 derivative of the HF service. The HF data is received using either a single or multiple channel demodulator which provides a composite RTCM output.

The Inmarsat version of Veripos, known as Veripos-I, was introduced in 1994 and consists of a consortium of companies which includes Subsea Offshore, Osiris and Topnav. The generally quoted accuracy of this system is better than 5 m to ranges in excess of 2000 km. Veripos-I uses up to fifteen reference stations connected to the system hub at Aberdeen. Correction latency within the system is better than 1 s. The hub sends the correction data to the Inmarsat uplink station for transmission to the satellite and retransmission to the survey vessels. The complete transmission and reception cycle is typically 3 s with an average age of correction of less than 5 s. The Inmarsat data can be received via a standard Inmarsat terminal and a dedicated Veripos-I demodulator. The correction data in RTCM format can then be fed directly to the user's GPS receiver or into a software package for QC analysis. If Inmarsat is not available, still sometimes the case on older smaller site survey operations, then Veripos can supply a 50 cm Marisat dome.

A further development by DSNP is a real-time kinematic technique called LRK (long range kinematic). LRK is the GPS processing technique based on the direct use of L1 and L2 carrier phase frequencies at 19 cm and 24 cm respectively as opposed to the technique of simple phase smoothing of code data at 300 m wavelength in conventional DGPS. RTK solves the problem of ambiguity of the phase data, also known as kinematic initialisation. The result is a system that is potentially ten times more accurate than conventional DGPS, that is, it is capable of centimetre accuracy. The system uses a receiver called Aquarius which has a maximum range of about 40 km.[21] This system is likely to be used for jacket positioning and for very precise engineering surveys rather than for seismic or site surveys.

Starfix

Fugro-Geoteam have also developed their own DGPS system known as Starfix. This system is entirely satellite based and was developed by a Fugro subsidiary, John Chance Associates. In its original form it dates from 1986 and became fully operational over North America in 1990. Starfix coverage is now world-wide and uses seven communications satellites with up-links from five earth stations.

Starfix performs the same basic tasks as Skyfix and Veripos and uses an intelligent user interface with an associated data display. It has a fairly extensive suite of data logging and processing facilities with the ability to interface with other pieces of navigation hardware. Starfix is able to handle multiple simultaneous position calculations. Positioning and heading filters are provided with Kalman filtering. For survey work there are two Fugro systems, namely Starfix-MN8 and Starfix-Spot. Starfix-MN8 uses the Inmarsat system for its differential corrections and Starfix-Spot transmits corrections via spot beam satellites.

Starfix can use D-GLONASS corrections in areas where GPS coverage is poor, particularly far northern latitudes.[22]

GPS tailbuoy positioning units

A typical tailbuoy positioning system is the Skytrak II relative GPS (RGPS) with the calculated GPS position of the vessel compared with the calculated GPS position for the tailbuoy and a range and bearing from the vessel generated. No differential corrections are necessary as both calculations use the same SVs. Positioning data is usually transferred to the vessel via a built-in UHF telemetry link.[23]

5.9 DGPS quality control

In considering DGPS quality control it is essential to mention that UKOOA guidelines for DGPS QC are incorporated into the software of all multi-reference systems.

In cases where differential stations are installed for a specific survey, great care should be taken to check the following.

- Antennae should have an uninterrupted view of the sky above the horizontal plane (zero degree elevation). The station site should be well away from electrical and communications interference. It is often the case that seagulls sit on the GPS antennae and seagull protection is essential in some locations.
- The station equipment will almost certainly be rack mounted and an air-conditioned building should be considered essential. If no building is available then a weatherproofed container may be suitable.
- Great care should be taken to minimise multipath effects caused by buildings in the vicinity of the installation. Cable runs between the antennae and receivers must be carefully planned, avoiding proximity to power lines and other electronic equipment.

Most contractors who offer DGPS use dedicated stations that are permanently installed and this form of operation is considered superior to differential stations that are mobilised for a particular operation.

Onboard the survey vessel facilities should exist to monitor the following parameters:

- Pseudo-range residuals for all satellites and the observation weight values used;
- Unit variance;
- Number of satellites in view and the number used in the position solution;
- Redundancy of least squares solution;

- Dilution of position values (HDOP, PDOP and VDOP) (horizontal, precision and vertical dilution of position);
- Latency of the differential correction;
- Position comparisons derived from different reference stations;
- Derived antenna height with respect to 'known' height;
- Monitor station information, especially position error measured at the monitor station;
- Maximum external reliability figure and the observation carrying it.

Each of these quality values should be continuously available in numerical form for every observed position fix. Statistical assessment, for example, means maximums and minimums and time series plots showing recent history (e.g. last 1–2 h) should also be available for display on demand. In addition, computer software used by survey contractors should include some quality assessment that relates to the following pair of factors:

- The a posteriori horizontal error ellipse as a precision measure;
- The largest horizontal position vector resulting from a marginally detectable error (MDE) as a measure of external reliability.

For quality control on most seismic operations it is assumed that the navigation computation is based on the principle of least squares and some form of continually filtered, on-line network adjustment process. In order to carry out rigorous quality control, a covariance matrix generated by the least squares computation should be used to generate test statistics and quality measures. The recommended test statistics are the w-test and the F-test (unit variance test). The w-test is used to detect bias errors. Those with a magnitude greater than 2.576 are highlighted for each position fix. The F-test is used to verify the model being used to account for errors in the DGPS observations (atmospheric refraction, clock offsets, earth motion). The average F-test value should be 1. Quality measures should include those discussed in 5.4, namely the error ellipse and external reliability.

Most site survey contracts state the minimum requirements for the operation of DGPS positioning systems. Typical survey parameter limits might include the following.[24]

- The reference point established for the GPS antenna should not be changed during the course of the survey.
- For real time quality control the Delta range DRNG should be < 7 m for 95 per cent of the shotpoints between different reference station solutions.
- The horizontal dilution of position (HDOP) should be < 2.5 for any shotpoint.
- The specifications for end of line statistical results might be as follows:
- PDOP (precision dilution of position) < 4 for 95 per cent of shotpoints;

Positioning systems 177

- HDOP (horizontal dilution of position) <2.5 for any shotpoint;
- Delta easting <5.0 m for 95 per cent of the shotpoints;
- Delta northing <5.0 m for 95 per cent of the shotpoints;
- Delta height <7.0 m for 95 per cent of the shotpoints;
- LPME (line of position mean error) <2.0 m for 95 per cent of the shotpoints.
- The specifications for times of poor satellite coverage or degraded signals might be as follows:
  - PDOP <4.0 m for 70 per cent of the shotpoints;
  - PDOP <6.0 m for 100 per cent of the shotpoints;
  - HDOP <2.5 m for 95 per cent of the shotpoints;
  - HDOP <3.0 m for 100 per cent of the shotpoints;
  - Delta easting <5.0 m for 95 per cent of the shotpoints;
  - Delta northing <5.0 m for 95 per cent of the shotpoints;
  - Delta range <7.0 m for 95 per cent of the shotpoints;
  - LPME <2.0 m for 95 per cent of the shotpoints.

Delta height is usually the difference in observed DGPS height and that calculated from known points such as the survey location. There will be geoidal undulations such as height of tide relative to mean sea level. Most marine survey computers allow for tidal range. It should also be noted that the antenna height above sea level and the geoidal undulation will change with location. Such factors may seem minor but can become significant on large area surveys. Delta height in times of poor satellite coverage should be an ellipsoid height compared to at least 6 h of 3D data. During data acquisition it is usual for the consulting engineer to log basic navigation data for quality control purposes. The precision dilution of position (PDOP) is usually logged on a line-by-line basis. The start- and end-of-line positions would be logged and compared with the survey pre-plots. The delta eastings and delta northings between two reference stations are also logged and compared. The shotpoint interval and the maximum off-line distance for a particular line would also be logged. The number of satellites available is an important statistic, also logged.

Positioning data is normally delivered in UKOOA P2/94 (raw data) and P1/90 (final data) formats on 3480 tapes.

5.10 Supervisory quality control systems

It is sometimes the case that the oil company will put a navigation quality control package onboard a site survey vessel and independently check every aspect of the contractor's positioning system. The following is regarded as a reasonable computer-based navigation package and any system designed to provide real-time quality control and post-processing quality control onboard the survey vessel should include the ability to measure position standard deviations, the F-test, the w-test, marginally detectable errors (MDEs), external reliability and range residuals. Other quality control features should include a three-dimensional error ellipse and a satellite sky plot. Before satellite positioning data can be accepted the computer program should check the validity of the ephemerides, the age of the satellite corrections and the elevation of the satellite. Observations can be rejected if the w-test (biases in the data) is unsatisfactory. The capability to reprocess logged raw data with different settings of elevation, reference station and satellite constellation should exist. A large set of statistics should be available in time series plots.

In exact terms, a pseudo-range should only be accepted for positioning after the following criteria have been satisfied:

- Validity of observation
- Availability of correction
- Elevation of satellite (user-defined mask)
- Availability of ephemerides
- Age of correction (age is user defined)
- Health of satellite (user controlled)
- Standby or disabled satellite by user
- Issue number of the ephemerides which are used in the calculation of the corrections

Quality control should be presented by means of standard deviation and height, error ellipses, residual errors, MDE and external reliability tests. The w-test, F-test and DOPs should also be available. Satellite signal-to-noise ratio, elevation, azimuth, number of satellites tracked and the number of satellites used for calculation are continuously available. Statistics should be available at the end of line which include the following:

- Residual errors
- Position standard deviation
- Marginally detectable errors (MDEs)
- F-test and w-test statistic
- External reliability
- PDOP, VDOP, HDOP and GDOP
- Delta easting and northing of the QPS derived position versus auxiliary positions
- Delta easting and northing of the static monitor position versus the theoretical position
- Delta easting and northing of the multi-reference position versus the single reference position
- Height

The basic operating philosophy of any onboard navigation QC package is provision of a high grade real-time monitoring function of the DGPS system positioning onboard the survey vessel. A typical onboard quality control package is the Exploration Consultants Limited (ECL) Q-Star system.[25]

Positioning systems 179

- Integrated twelve-channel Motorola GPS receiver, with a minimum of twelve parallel channels;
- Optional Trimble 4000 receiver input;
- 486 PC and monitor and eight-port TCL card configuration;
- Software with full UKOOA testing and comparison displays;
- DGPS monitor station and data transmission facility.

This system applies the pseudo-ranges derived from the Motorola receiver to the contractor-supplied differential link and computes a weighted least squares position solution within the software. The use of a separate receiver enables a performance comparison with the primary in-use receiver, while the preferred common use of the contractor antenna, wherever feasible, will indicate the relevant multipath or signal attenuation to the QC system. The QC software then performs a comprehensive statistical analysis on all observables to assess the validity of all received data being applied in the least squares solution. The final QC position is then compared with that of the contractor-derived DGPS position (additional positions derived from secondary DGPS or radio navigation aids may also be inputted for direct comparison). All data can be recorded for playback and/or statistical print-out.

During data acquisition four bar graphs should always be displayed. These represent the number of satellites available, the age and parity of the differential link, the magnitude of the major axis of the 95 per cent 2D error ellipse and the radial difference between the QC and the contractor position. Each of these graphs should be configured with a warning limit which, if exceeded, sets off audio alarms and turns the bar graph from green to red. This configuration of graphs enables a quick reference to the quality of performance against defined benchmarks.

The QC analyses chart should give a comprehensive statistical testing tool. The tests are presented numerically and as a time series. The test incorporates F and w testing, data snooping and marginally detectable errors (MDE).

The map display is a fully scaleable map with waypoint marking and navigation information. The QC and contractor positions are displayed for comparison and a 95 per cent error ellipse centred on the QC position is also presented.

The differential link monitor is a full graphical representation of the contractor-supplied differential link. This includes pseudo-range corrections, range rate corrections, age of corrections and any other information that is transmitted via the RTCM. Up to six reference stations may be monitored and displayed with individual satellite data analysis. Two separate differential modules can be monitored simultaneously.

The multiple reference station solution is a definable multiple reference station solution which can be selected and computed with combined data displayed on the differential link module.

The DGPS performance graphs and read-outs consist of a number of time series performance graphs which include satellite signal strength and tracking,

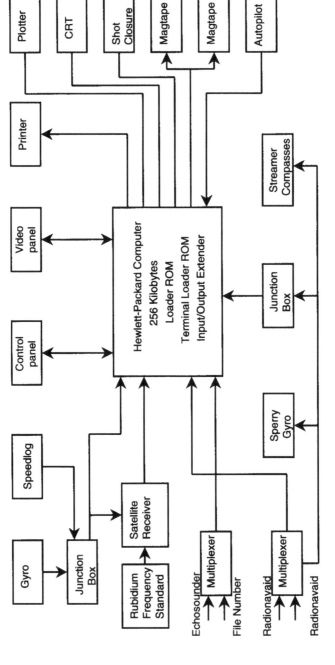

Figure 5.4 Older type of integrated navigation system.

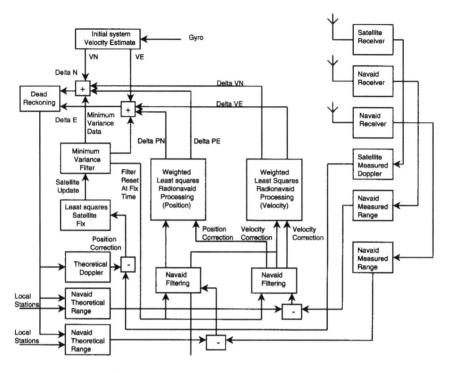

Figure 5.5 Integrated navigation program structure.

dilution of position (DOP), position comparisons in northings and eastings, height comparisons and vector/scatter plot diagrams for static verifications.

Satellite planning diagrams are skyplots for real time and on a 24 h basis are available for satellite constellation analysis and planning. The 24 h skyplot is complemented by a satellite visibility bar chart.

5.11 Integrated navigation systems

Integrated navigation systems have already been mentioned in the context of fire control for seismic gun arrays. Fully integrated navigation systems of the type used for 2D and 3D seismic surveys are used only occasionally for site surveys. However, they are a feature of some site survey operations, so are included here for historical completeness. Figures 5.4 and 5.5 show an older type of integrated navigation system and the integrated navigation program structure that might be associated with it.[26] Another older type of integrated navigation system used for low cost 2D and high resolution surveys is the Geofix Scope 111. A typical integrated navigation system for site surveys will be computer based and there is a vast range of computers available. The differential global positioning system (DGPS) is integrated to the

182 High resolution site surveys

navigation computer and the position calculated. This is usually performed by the pseudo-range technique or through some other type of navigation processing. This is inevitably a computer program that integrates the DGPS range data to the integrated navigation system (INS). In the computer, typically three slots are used for latitude computations in degrees, minutes and seconds. Three are used for longitude computations in degrees, minutes and seconds while a further two slots give the position read-out in eastings and northings.

Position calculations are made by independently weighted least squares adjustments performed on the DGPS data. A quality position derivation is calculated and displayed as both the residual errors from the least squares routine and as a standard deviation.

The most modern system in use today is the Concept Systems SPECTRA, suitable for 3D, 2D and high resolution site surveys. This system is modular and comprises various nodes. The system is based on an expandable network of UNIX workstations. In its most basic form it provides source and streamer positioning for the most simple high resolution surveys. At its most complex it can handle multi-streamer, multi-source and multi-vessel operations, allied to the most sophisticated streamer acoustic positioning systems used for 3D seismic surveys. There are up to twenty nodes that define every level of navigation activity, including two for quality control. The number of nodes used will be determined by the sophistication of the survey and its requirements.[27] Figures 5.6 and 5.7 show schematics of the SPECTRA system.

One contractor has used the old Geophysical Service International (GSI) integrated navigation system, known as the configurable marine system (CMS). This system takes a number of navigation sensors and gives a position based on them. The centre of the system was a Hewlett-Packard 980B computer. In common with the Geofix Scope III system, pseudo-Argo, Hyperfix or Syledis ranges were required for DGPS position fixing.

Box 5.4
Integrated navigation systems

Some people will tell you that integrated navigation systems are not used on site survey operations, merely a small computer that gives the distance run every 100 m or so. This is partly true but the site survey industry has used so much 'cast-off' seismic equipment from the big contractors over the years that you can never guarantee that particular systems will not turn up on a site survey operation. One of the big changes in the industry in recent years has been the adoption of DGPS for nearly all survey work. Everyone works to the same specifications, usually those of the UKOOA, and there is no particular reason why integrated navigation systems should not be used on site surveys, hence the descriptions given here.

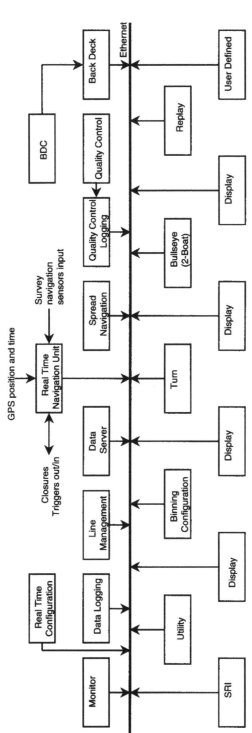

Figure 5.6 Concept Spectra integrated navigation system.

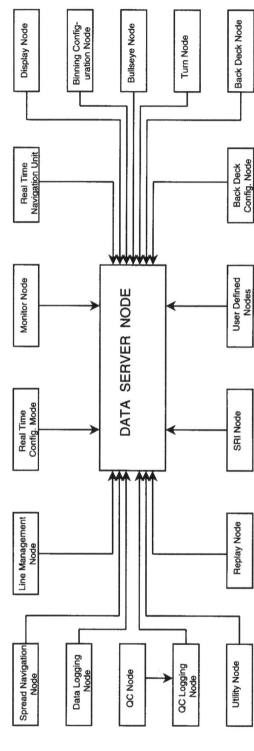

Figure 5.7 Spectra nodal networks.

Positioning systems 185

Finally, most integrated navigation systems have a gyro-compass and echosounder integrated. A Doppler sonar may also be included but this is virtually unknown for site surveys. Figure 5.8 shows a complete site survey system with all the positioning, seismic acquisition, analogue acquisition and interfaces in place.

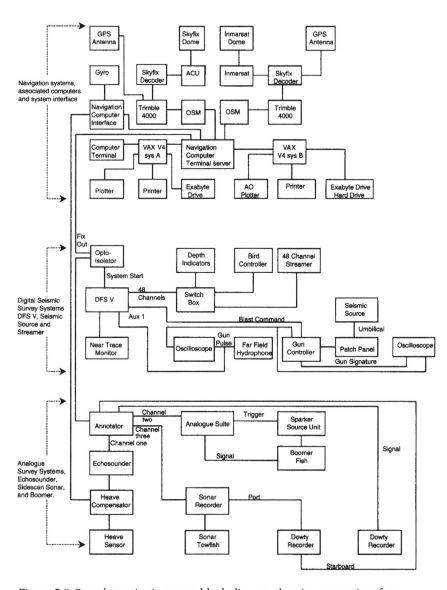

Figure 5.8 Complete seismic system block diagram showing system interfaces.

186 *High resolution site surveys*

5.12 Acoustic positioning systems

In recent years acoustic positioning systems have proliferated. Three-dimensional surveys now use very large acoustic networks to accurately position up to twelve streamers and four sources. For site surveys acoustic positioning is normally associated with sonar towfish positioning and streamer tailbuoy positioning. Very deep-tow sidescan sonar and swath bathymetry systems need accurately positioned towfish, particularly in the case of systems where up to 10,000 m of cable may be deployed. There is a great deal of interest in remote operated vehicles (ROVs) and autonomous underwater vehicles (AUVs) for deep sonar mapping. It is possible that the first 3D site surveys may be conducted in the near future so acoustic positioning may become relatively more important in future than was previously the case.

Acoustic positioning has two different techniques and three separate frequency bands in use. In the first technique an array of acoustic transponders will be placed on the seabed and referenced to each other. The grid is then referenced to the geodetic datum with DGPS. In the second, the acoustic network on a streamer, or the sonar towfish is referenced to a point onboard the survey vessel. This point can be a single point such as a fixed transducer or perhaps a port–starboard baseline with transponders at either end.[28]

In terms of frequency the three basic categories of system include long baseline (LBL) systems operating at 8–16 kHz, medium frequency (MF) systems at 18–40 kHz and extra high frequency (EHF) systems, which operate above 40 kHz. In terms of usage, LF systems are usually used for deep navigation and research while MF systems are used for sub-sea vehicle tracking and streamer positioning. EHF-band systems are used for research, survey and meteorology.

The LBL system measures ranges to transponders that are at known points on the sea floor, usually in an array. The system measures ranges from these known points and calculates the surface transponder position. The advantage of such a system is that being sea-floor referenced it does not need a gyro. Regardless of vessel roll, pitch, heave or yaw the seabed acoustic grid is a fixed reference baseline. Baseline lengths in such arrays are usually in the order 100–6000 m. A radius of 40 per cent of the water depth is often quoted as an effective baseline length. The major errors associated with LBL systems include range detection (timing) errors, field calibration errors and ray bending problems as slant ranges increase. The chief disadvantage of LBL systems is the lengthy calibration procedures associated with laying a sea-floor network of transponders.

With SBL systems a range and direction is established with a single sea-bed transponder. The survey vessel now requires a gyro, properly calibrated for survey purposes, and a device for measuring vessel pitch and roll. Vessel reference units (VRUs) are in widespread use. Motion reference units (MRUs) have the additional capability of measuring vessel heave. At least four transducers will be mounted onboard the survey vessel, three for triangulation

Positioning systems 187

with the seabed transponder and one spare. Errors in SBL systems include timing errors and platform motion errors. The main advantage of SBL systems is a much more rapid calibration and a faster data update. The range of SBL systems is rather limited compared to LBL systems, typically 2 km.

Variations on the two categories of systems include supershort or ultrashort baseline (SSBL or USBL). These systems measure a range and angle, horizontal and vertical to the sea-floor transponder and need a VRU as a mandatory requirement. Other combination systems include LBL and SS/USBL systems.

Typically an ultra-short baseline (USBL) acoustic positioning system would be used to monitor the position of the sidescan sonar towfish and the streamer tailbuoy position. A signal is transmitted from the hydrophone assembly onboard the survey vessel and received by a beacon which transmits a signal back to the hydrophone assembly. The time of travel of an acoustic signal through the water is proportional to distance of travel. The very short baseline method is able to pinpoint the position of the single beacon in three-dimensional space.

Acoustic errors

For an acoustic system to work properly a signal-to-noise ratio better than 20 dB is a basic requirement. Survey vessel noise from thrusters and propellers is the principal source of noise in acoustic systems. Survey vessels produce noise at around 125 to 127 dB/µPa per hertz, with exceptionally noisy thrusters producing as much as 140 dB. The noise spectrum level usually peaks at around 100–1000 Hz. Other sources of acoustic noise likely to be encountered include vibration and generator noise, seismic activity, trenching, piling, etc. Background environmental noise in the MF band is typically less than 80 dB/µPa per Hz and can be largely ignored. The main considerations in evaluating likely sources of acoustic noise are the distance from the transducer to the propellers or thrusters, the transponder source level, the transponder directivity and the transponder bandwidth.

Other sources of error include air bubbles, ray bending and multi-path interference. Microscopic air bubbles can be excited by the acoustic transmission and this will scatter and distort the acoustic signal. This can be detected by the presence of second harmonics (frequency doubling) in the echoes. Micro-bubbles can also be created by rapid vessel movement (pitch, heave, roll). Ray bending is caused by sharp changes in temperature and salinity at various depths which cause sharp velocity changes in the acoustic energy. Acoustic velocities increase at a rate of about 1/60 m per second per metre of depth increase. A decrease in temperature decreases the transmission speed by about 3 m/s per °C. There is also a decrease in velocity with increase in salinity. These velocity changes result in the refraction of the acoustic wave front and ray bending. The worst ray-bending problem usually occurs when the horizontal distances exceed 2–3 times the water depth. A temperature/salinity dip is usually part of the acoustic calibration.

Multi-path interference occurs when refracted or reflected signals arrive at a receiver at the same time as direct signals. Acoustic signals can be reflected off parts of the vessel, the sea surface, seabed or sub-sea structures, shoals of fish, plankton, etc. Normally this reflected signal will arrive at the receiver well after the direct signal and can be rejected. Problems occur when the indirect path is only just longer than the direct path. This tends to happen in shallow water and at longer acoustic ranges.

In acoustic positioning, accuracy is considered to be the absolute nearness to the truth of a series of observations, while repeatability is the relative grouping of a series of observations without regard to the absolute nearness to the truth. Accuracy is important when returning to a wellhead and for surveying. Repeatability is usually stated in percentage RMS of slant ranges. Distance root mean square (DRMS) is the RMS value of the distances of individual scatter points from the centre of a range circle.

Acoustic calibrations

A first requirement for acoustic calibrations is to check the gyro alignment and to perform a VRU calibration. These will be conducted in port before the start of survey. A gyro calibration involves establishing a baseline at the dockside and measuring offsets from this baseline, fore and aft, to the centreline of the vessel. The difference between the steady reading and the bearing of the baseline is the misalignment of the gyro. While in port the transducer alignment relative to the vessel's head, offset from the transducer(s) to the vessel's reference point(s), and VRU settings for list and trim should be checked.

Sound velocities should be established by temperature/salinity checks at the survey area. Once the transponders have been laid on the seabed, baseline crossings will be performed and the transponders 'boxed in' to an absolute position by taking slant range measurements. Once the position of two transponders relative to a DGPS position is known, a second 'box-in' overcomes any gyro errors.

With SSBL/USBL systems the rotational errors about three axes are independent of each other. Errors in rotation about the z-axis are considered to be alignment errors, while y-axis errors are caused by vessel pitch and x-axis errors are caused by vessel roll.

Before the start of survey a USBL calibration would usually be undertaken. This typically comprises a static and dynamic calibration. A transponder would be deployed on the seabed. After determination of the mean water column velocity the static phase of the calibration involves the vessel remaining stationary above the transponder to allow accurate determination of the depth of the beacon. This is referred to as the Z value. Typical calibration figures might be as follows.[29]

- Mean velocity of sound: 1503.3 m/s
- Mean depth by echosounder: 87.8 m
- Mean observed Z value: 87.4 m (SD = 0.5)

Positioning systems 189

Once an accurate depth measurement has been made the transponder is boxed in by a series of lines in both a clockwise and anticlockwise direction. This allows any alignment and scale error to be determined and corrected for, during survey operations. Again the figures given here may be taken as typical.

- x offset: -0.47 m
- y offset: -1.20 m
- z offset: 5.1 m
- Alignment: $-1.71°$
- Pitch: 0.0
- Roll: 0.0
- Scale: 0.9999997

Systems in use

Sonardyne now offer their seismic streamer and source positioning system, the Sonardyne integrated positioning system Mk 2 (SIPS2). This system is designed principally for big, multi-streamer 3D seismic operations. By judicious use of an USBL acoustic system Sonardyne now have a system capable of positioning a towfish in water depths of 3500 m, with a cable layback of 10,000 m. This is illustrated in Figures 5.9 and 5.10. The towfish operates about 100 m above the seabed with slant ranges too long for normal acoustic positioning with MF frequencies. A second survey vessel (chaseboat) is fitted with a USBL positioning system.[30] The USBL is then synchronised to an external trigger. The towing vessel initiates a recording sequence by sending DGPS-derived signals to the towfish which sends out an acoustic pulse. Signals from the towfish transponder are then received by the

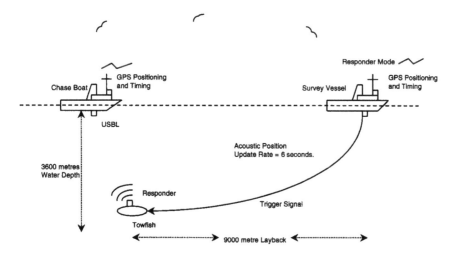

Figure 5.9 Acoustic positioning of a sidescan sonar towfish.

190   High resolution site surveys

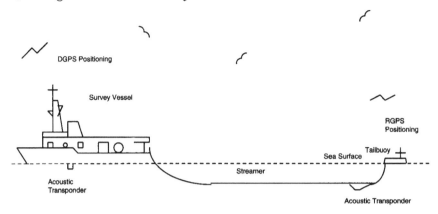

Figure 5.10 Acoustic positioning of a seismic streamer.

chaseboat and the position of the towfish is computed and displayed. This data is combined with the chaseboat's DGPS data and is then telemetered to the survey vessel. The update accuracy of such a system is about 3 s.

The Simrad high precision acoustic positioning (HiPAP) is designed to operate with ROVs, towed bodies or fixed transponders.[31]

References

1. Parkinson, R. (1980) 'Data quality control on survey navigation systems', Paper presented to the Petroleum Institute Training Board.
2. Sleigh, W.H. (1988) *Aircraft for Airborne Radar Development*, HMSO, London, p. 7.
3. McQuillan and Ardus (1977) *Exploring the Geology of Shelf Seas*, Graham and Trotman, London; p. 13.
4. Parkinson, R. (1978) 'Onboard manifestations of noise and errors detectable in the survey situation', *Norske Petroleum Forenung*.
5. Leick, A. (1990) *GPS Satellite Surveying*, John Wiley and Sons, Chichester, p. 1. (Long excerpt reproduced by courtesy of the publishers.)
6. 'GPS/GLONASS accuracy', *International Ocean Systems Design*, September/October 1998.
7. 'Ashtech GG-24 receiver', *International Ocean Systems Design*, September/October 1998.
8. 'Evaluation of GLONASS', *The Hydrographic Journal*, April 1998.
9. Leick, op. cit., pp. 58–59. (By kind permission of John Wiley and Sons.)
10. Leick, op. cit., p. 60. (By kind permission of John Wiley and Sons.)
11. Leick, op. cit., pp. 60–61. (By kind permission of John Wiley and Sons.)
12. Stansell, T., Jr (1973) *Accuracy of Offshore Navigation Systems*, Magnavox Research Laboratories.
13. *Geodetic Definitions and Datums*. Geco Training Document associated with The Northstar Integrated Navigation System.
14. Ashkenazi, V. (1978) 'A short introduction to classical geometrical geodesy', *Seminar on Satellite–Doppler Positioning*, Nottingham University, 1978.

15. Spradley, L.H. (1987) 'New co-ordinate system changes offshore grids', *Ocean Industry*, January 1987.
16. Loweth, R.P. (1997) *Manual of Offshore Surveying for Geoscientists and Engineers*, Chapman and Hall, London, p. 133.
17. Langley, R.B. (1998) RTK GPS description taken from the Internet. Other subsidiary references include an LRK article in the *Hydro International Journal* (April 1998) and an article in the *Hydrographic Journal* (1998), also an *Aquarius RTK Journal* article.
18. *Admiralty List of Radio Signals*, Volume 8, 1998/99, Satellite Navigation Systems, Taunton.
19. 'Introduction to quality assessment and statistical testing', DGPS Guidelines Section 2, p. 24.
20. Skyfix description. Information pack provided by Racal Survey.
21. Veripos description. Information pack provided by Veripos. 'DSNP DGPS', *Hydro Magazine*, May/June 1998.
22. Starfix description. Information pack provided by Fugro-UDI.
23. Taken from a survey report written by the author.
24. Contract specifications given here represent a fair summation of five sets of specifications used by the author in the mid 1990s, for 2D seismic and site surveys.
25. The description given here is based on the ECL Q-Star system, by kind permission of Exploration Consultants Limited.
26. The drawing in Figure 5.4 is essentially the Magnavox integrated system. The description of integrated navigation systems is taken from numerous old QC reports on seismic contractors.
27. Description of the Concept SPECTRA system and the block schematic diagrams reproduced by kind permission of Concept Systems.
28. 'Basic principles and use of hydroacoustic position reference systems in the offshore environment', IMCA Document, IMCA M 151, April 1999.
29. Oceano Satellite Integrated Acoustic Positioning System. This system was relatively widely used for site survey work. The author has referred to three site survey reports for the description of calibration techniques.
30. 'Acoustic fish positioning. Sonardyne sensor', *News from Sonardyne*, Spring 1995.
31. 'Autonomous underwater vessels', *Offshore Engineer*, May 1999.

6 Safety

6.1 Safety reviews

Most oil companies conduct safety audits on vessels working within the confines of their oilfields and in areas covered by the exploration licences granted to them. Paradoxically, site surveys often fall outside the normal safety rules operated by oil companies. In consequence they tend to be conducted without safety audits and with a minimum of attention paid to safety and safety-related matters. As a general rule it is strongly recommended that a full safety audit be conducted on a vessel that is newly mobilised for site survey work, particularly if the vessel in question has not performed site surveys before. Beyond this recommendation the decision to conduct a full safety audit lies with the oil company but regular safety audits are recommended. During the course of a survey the consultant engineer assigned to a particular survey will comment on safety and safety-related matters and this will form a separate part of his report. In its broadest sense the word safety covers all aspects of health and environmental matters, in addition to considerations of the survey ship and its equipment.

Safety audits and reviews are governed by a number of international regulations which include The International Convention for the Safety of Life at Sea 1992, known as SOLAS 1992, and subsequent amendments. There is also a 1992 SOLAS Consolidated Edition of the International Marine Organisation (IMO) regulations. The UK Department of Trade (then Transport, now the Maritime and Coastguard Agency) publish a Code of Safe Working Practices for Merchant Seamen and state the standard commonly met by seismic survey vessels operating in the North Sea. The World Health Organisation (WHO) publish an International Medical Guide for Ships. In addition The International Association of Geophysical Contractors (IAGC) publish a marine geophysical operations safety manual. E and P (Exploration and Production) forum rules usually apply. At the time of writing Exploration and Production has been renamed Oil and Gas Production schedules. The E and P (Oil and Gas) rules likely to affect site surveys include but are not limited to the following:[1]

- E and P Forum 1991b. Substance Abuse Management Strategies
- E and P Forum 1993a. Aircraft Management Guide

- E and P Forum 1993b. Exploration and Production Waste Management Guidelines
- E and P Forum 1993c. Guidelines on Permit to Work (PTW) Systems
- E and P Forum 1994a. Guidelines for the Development and Application of Health, Safety and Environmental Systems
- E and P Forum 1994b. Generic Hazard Register for Geophysical Operations
- E and P Forum 1995. Guidelines for the Use of Small Boats in Marine Geophysical Operations
- E and P Forum 1993. IAGC Safety Training Guidelines for Geophysical Personnel
- E and P Forum 1993. Safety Training Guidelines for Geophysical Personnel
- E and P Forum 1993. Oil and Gas Exploration and Production in Arctic and Sub-Arctic Regions. Guidelines for Environmental Protection
- IAGC 1991b. Marine Geophysical Safety Manual, Eighth edition
- IAGC 1994. Environmental Guidelines for World-wide Geophysical Operations
- IMO 1978. Marpol Annex IV Regulations for the Prevention of Pollution by Sewage from Ships

---

Box 6.1
Safety nomenclature and associated paperwork
In attempting to write a chapter on safety the multiplicity of documents and handbooks available became thoroughly bewildering. Many of the operating agencies concerned with safety appear to have changed their names at least three times during the 1990s. The Department of Trade became the Department of Transport and then the Maritime and Coastguard Agency. Exploration and Production became Oil and Gas Production, etc. The author makes no apology for not being completely up to date; any published work will always be, to some extent, out of date. This chapter suggests an approach to safety that starts with a safety audit, followed by a detailed safety regime for the survey vessel and its equipment. It finishes with safety as an aspect of personnel competence.

Many oil companies specify their own safety regime. For example, Shell have a 1985 document concerning enhanced safety management. This document defines eleven safety management elements (ESM principles) used in the Shell group of companies prior to the introduction of HSE management systems. Shell document SHSEC 1989 explains the eleven ESM principles in detail.

Safety reviews usually cover the general safety arrangements, working conditions and practices and a general review of standards of safety and safety awareness.

Existing UK legislation, regulations and worldwide guidelines that relate to site surveys include the following:[2]

- Coast Protection Act 1949
- Continental Shelf Act 1964
- Mineral Workings (Offshore Installations) Act 1971
- Petroleum and Submarine Pipelines Act 1975
- The Health and Safety at Work Act 1974
- Offshore Safety Act 1992
- Petroleum Act (Production) (Seaward Areas) Regulations 1988
- The Offshore Installations (Safety Case) Regulations 1992
- The Offshore Installations and Wells (Design and Construction, etc.) Regulations 1996
- Application for Consent to Drill Exploration, Appraisal and Development Wells 1996
- Record and Sample Requirements for Surveys and Wells 1996
- Notification of Geophysical Surveys 1996
- Guidelines for the Use of Differential GPS in Offshore Surveying UKOOA 1994
- Environmental Guidelines for Exploration Operations in Near-Shore and Sensitive Areas 1995
- Guidelines for Minimising Acoustic Disturbance to Small Cetaceans, D of E 1995
- Offshore Installations: Guidance on Design Construction and Certification 1990
- A Guide to the Offshore Installations (Safety Case) Regulations 1992
- New Guidance on the Coast Protection Act 1995
- Environmental Guidelines for Worldwide Geophysical Operations IAGC, 1996
- Procedures Relating to Notification of Vessels Intending to Anchor in the Vicinity of Pipelines or other Subsea Installations 1986

The regulation of surveys will vary from country to country but the UK regulations given here can be taken as typical. From the foregoing it is apparent that the 1990s have seen a proliferation of safety-related legislation and this is unlikely to decrease with the passing years.

6.2 Documentation

As part of a full safety audit the vessel documentation should be reviewed in some detail. The port of registration and classification authority should be known and approved by the oil company. The certificate of registry, the international tonnage certificate, the annual loadline certificate, the stability booklet, the minimum safe manning document and the certificates of competence and training for the vessel master, officers and ratings should be inspected. The international oil pollution prevention certificate, vessel liferaft certification, de-ratting certificate and safety radio certificate should all be valid for the period of the survey. A safety equipment certificate should cover all these points in one document. Hull and main engine survey certification should also be checked. Letters and certificates relating to vessel stability and operating restrictions should be made freely available by the vessel owners.

The gross registered tonnage of the survey vessel should be checked and it should be noted that cargo ships of less than 500 gross registered tons are exempt from many of the SOLAS requirements. There has been a tendency of certain seismic contractors to register their vessels at less than 500 gross registered tons by judicious use of the bureaucratic ratings for tonnage operated by some foreign authorities. For example, some registry authorities define gross registered tonnage in terms of enclosed space up to the main deck level, ignoring the vessel's conversion to seismic work and large extensions above main deck level. Such bureaucratic economies with the actual truth about the vessel's size do not usually prevent a vessel working for the oil company, but they should be known about and reported upon.[3]

6.3 Health and safety policy

All contractors should have a health and safety policy. It is now the trend to implement this policy by formal management systems. Onboard the survey vessel all safety procedures should be written as a formal safety manual. This Health, Safety and Environment Management System (HSE-MS) should control all aspects of field work. There should be at least three copies of the safety manual available onboard the survey vessel. This document should be reviewed by the oil company or other operating authority as a matter of normal site survey supervision.

When reviewing the contractor's safety policy the following should be considered:

- Pre-tender contractor health, safety and environment assessment;
- Health, safety and environment clauses in the contract;
- Pre-award briefings;
- Contractor's health, safety and environment training;
- Contractor's health, safety and management workshops;
- Contractor's health, safety and environment systems, including hazard identification and working procedures;
- QC supervision to include health, safety and environment;
- On-site health, safety and environment audits;
- Logging of safety incidents followed by analysis and feedback;
- Logging and analysis of performance indicators.

Vessel health, safety and environmental audits should be undertaken both before and during operations.[4]

6.4 Incident and accident reporting policy

Accident reporting procedures should require all accidents to be reported to the shore management. All onboard accidents should be immediately reported to the onboard oil company consulting engineer. All accident reporting should include a detailed written account of how the accident took place, where and when it took place, names of all witnesses to the accident and measures to prevent recurrence of the accident. All lost time incidents, near misses and dangerous occurrences should be notified to the oil company and to the management. Figures 6.1, 6.2 and 6.3 show a hazard observation chart, an accident/incident reporting flowchart and an outline geophysical response organisation flowchart.[5] United Kingdom legislation applies to all foreign flag vessels while in UK ports or UK territorial waters. This means that RIDDOR – The Reporting of Injuries, Diseases and Dangerous Occurrences Regulations 1985 applies. This procedure requires immediate notification to the enforcing authority of serious injuries and of certain dangerous occurrences. The HSE 11 (rev) booklet explaining the regulations and the F2508 form should be held onboard for RIDDOR reporting purposes.

6.5 Cranes and lifting gear

Most seismic and site survey vessels are equipped with hydraulic cranes, used for handling heavy equipment and for the deployment/recovery of survey gear. The Safe Weight Load (SWL) markings on the jib should be inspected on all site surveys and inadequate or obscured markings should be pointed out immediately. There should be an annual inspection of all cranes and a 4-yearly complete inspection. The survey company should hold onboard a certificate of periodic inspection by an independent competent person. The jib sheave should be free and the crane checked for defects as part of the vessel mobilisation. The deck surrounding the crane should be grease and obstruction free and crane working areas should be designated hard hat and safety boot areas. Cranes, winches and hydraulic equipment should be operated by or under the supervision of trained personnel. Personal safety equipment must be worn by all personnel handling cargo or working in the vicinity of the crane.[6]

6.6 Working conditions and working practices

Deck areas should always be treated as hazardous. On the upper decks all boarding gangways should be rigged with a safety net. Upper decks and walkways should be clear of obstruction and preferably treated with a suitable

Safety 197

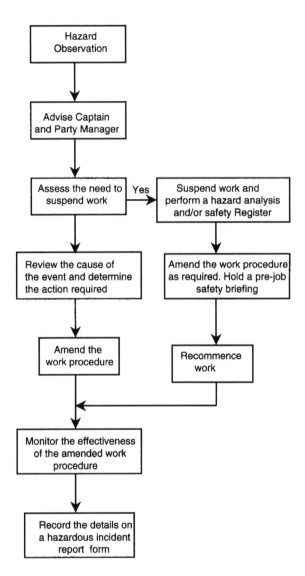

Figure 6.1 Hazard observation flowchart.

non-slip surface. Approaches to ladders and stairs should be at least 16 in. wide, unobstructed and also treated with non-skid material. Fixed ladders, landings and cages should be inspected frequently and properly maintained. Inside compartments, instrument rooms and stores, compressor and sparker rooms and workshops should be roomy with easy physical access. Access and egress points, walkways, etc. should be well lighted with directional arrows indicating emergency exit points. Obstructions, danger areas and areas of

198 *High resolution site surveys*

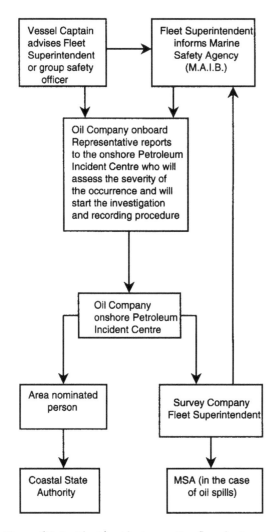

Figure 6.2 Accident/incident reporting flowchart.

high noise levels should be properly identified. Ropes should be kept free of contamination by chemicals such as rust removers and paint strippers. Persons working aloft, outboard or below decks or in any area where a risk of falling more than 2 m exists, should wear a safety harness attached to a lifeline.

Watertight doors should be closed and latched at sea. Door and hatch gaskets should be kept clean and checked regularly. Sand should be used on slippery areas caused by ice, snow or rain. Oil spillages and grease should be cleaned off decks as soon as practicable. In rough weather it may be necessary to rig lifelines across open decks. Permanent fittings that cause

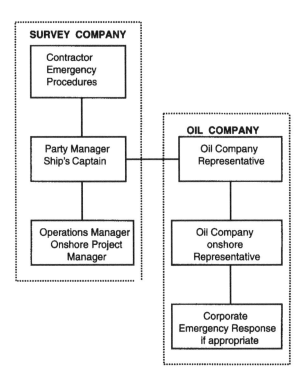

Figure 6.3 Outline geophysical response organisation.

obstruction such as eye plates, lashing points and projections should be painted a conspicuous colour in contrast to the background. Machinery guards should be kept in place and in good condition.

It is particularly the case with site surveys that the vessels are internally cramped and a great deal of survey equipment will be stored on the upper deck along with oil drums and assorted ship's gear. A careful check should be made that all survey equipment is properly secured, that all oil drums are clearly marked with the contents identified and that no grease or oil is deposited on the decks. All open edges should be protected and handrails and safety chains should be in good order. Paint should be held in designated lockers with all inflammable liquids kept in closed containers. Outboard motor fuel (gasoline/petrol) should be stored in ventilated lockers on an open deck well away from any source of ignition. Inflammable stores and accesses to the cable deck should be marked with 'no smoking' notices. Most survey vessels hold small quantities of hazardous substances such as solvents and adhesives. US and UK regulations require that data sheets for such substances should be held onboard to provide information on the safe method of storage and use and the measures to be taken should inhalation, ingestion or splashing occur.

If gas welding gear is used the set should be fitted with a flashback arrestor on the acetylene line. It is best practice to fit a flashback arrestor to the oxygen line as well. It is increasingly the case that hot work such as cutting and welding should be subject to a formal 'permit to work' system. 'Permit to work' systems also apply when working at height on antennae and radar installations. 'Permits to work' may also include entry to close spaces and work performed over the side of the vessel.

In the ship's engine room all exhaust pipes and fittings should be heat insulated. High noise areas, where ear protection is compulsory, should be fitted with visual alarms in case an audible alarm is shielded by the ear protectors. Waste oil should not be allowed to accumulate in the bilges and any accumulation should be disposed of in accordance with oil pollution regulations. Bilges should be painted a light colour and kept clean and well illuminated. Bilges should be kept clear of rubbish so that mud boxes are not blocked and the bilges can be pumped easily and quickly. Remote controls, fire valves for stopping machinery pumps or for operating oil settling tanks should be tested on a regular basis. Safety warning notices should be clearly displayed.

Procedures for launching and recovery of survey gear should be covered by formal written procedures contained within the contractor's safety programme manual. During upper deck crane work, survey equipment deployment and recovery and operations such as coring all personnel involved should wear hard hats and steel-capped safety shoes/boots. Personnel working to the rear of the cable deployment deck should wear a buoyancy aid and lifeline harness. Designated safety footwear areas should be clearly marked, typically cable and gun/sparker deployment areas and crane working areas. Safety footwear and hard hats should be available on a free issue basis to all onboard personnel.

It is a common occurrence that unsafe acts are committed by shore personnel engaged in maintenance and repair onboard a survey vessel in port. All visitors onboard a survey vessel should report to an onboard co-ordinator designated for this task while the vessel is in port. A briefing sheet should be available, drawing attention to safety regulations in force on the vessel and any precautions that should be observed for the particular task in hand.

Any grinding wheel in the sparker or gun shack should be free of chips or cracks. A transparent hand guard should be in place. Safety goggles should be prominently displayed above the grinding wheel and a first aid kit should be nearby. Eyewash kits should also be available and a notice stating that eye protection is compulsory when the grinding wheel is in use. Chuck guards should be fitted to pillar drills. All machinery guards should be kept in place and in good condition. Workshop machinery should ideally be fitted with emergency stop buttons.

The galley area should be subjected to close inspection. Meat and salad vegetables should be stored on separate trays and preferably in separate refrigerators. There should be no smoking in the galley or instrument room. Persons employed in the preparation, cooking or serving of food and drink, or in the handling of eating and drinking utensils, must have undergone a medical examination and fulfil the necessary health requirements. The galley personnel should wear clean protective clothing with long sleeves and hair covering. A large proportion of injuries to catering staff are caused by slips and falls that result from the wearing of flip-flops, sandals or plimsolls. Such footwear affords no protection to the feet from burns and scalds if hot or boiling liquids are spilt. Suitable footwear, preferably with non-slip soles, should be worn by galley staff. The galley should be bleach washed at least once per day. It is 'best practice' if a steam steriliser is used for all freshly washed cutlery and crockery. Cockroach infestations should be treated vigorously and noted in the client representative's report. Ship amenities should include a designated mess room, a recreation room, laundry room and adequate accommodation based around a mixture of one- and two-berth cabins.[7]

All refrigerated room doors should be fitted with a means of opening the door from inside and an alarm bell to warn those outside of person(s) trapped inside. Non-slip flooring should be fitted outside refrigeration rooms.

6.7 Fire protection equipment

The vessel engine room and machinery spaces should be covered by a fire detection system with a display console fitted on the bridge. Independent battery-powered smoke detectors are the minimum requirement in the accommodation areas. A vessel fire main should be connected to hydrants provided with standard hose lengths. These should all be checked during the mobilisation phase of the survey. All extinguishers should have their contents clearly marked in English and the prevailing national language of the crew. Some form of foam hydrant should cover the compressor room and cable reel. As an alternative, fixed water deluge systems could cover these areas. Portable dry powder fire extinguishers are usually installed around the ship and a safety audit should check the expiry dates on all extinguishers. The galley area should be fitted with a foam or Halon extinguisher and a fire blanket. The engine room should be fitted with a water hydrant and chemical dry powder extinguishers. It is recommended that machinery spaces should be covered by CO_2 or Halon systems, though it is pointed out that Halon is a greenhouse gas and is being phased out. Breathing apparatus should be held at two locations, as a basic minimum. There should be adequate spare air bottles for training purposes. A fire proximity suit should be held onboard equipped with a harness and lifeline.[8]

If it is necessary to disable the ship's fire alarm system in a specific work area permission must be obtained from the officer of the watch.

6.8 Vessel safety and survival equipment

Liferafts should be carried sufficient to carry approximately twice the number of persons onboard. All liferaft stowages should be fitted with hydrostatic releases. It is usually the case that small ten-person liferafts are fitted with quick release mechanisms and larger twenty-person varieties are fitted with float free stowages. Liferafts should be identified by number and allocation of personnel posted in individual cabins. If a totally enclosed motor-propelled survival craft (TEMPSC) is carried its capacity should be equal to the total number of people onboard. In addition the liferafts should have an equal personnel carrying capacity. Checks should be conducted on the TEMPSC to show that the craft is regularly lowered to the water and its engine started. It should be established that the required survival equipment is onboard.

It is a SOLAS requirement that muster and embarkation stations are provided with lighting from an emergency source of electrical power. Liferaft launching instructions should be prominently displayed and this matter explained during abandonment drills. Rope ladders should be kept for boarding liferafts from the main deck level, particularly if the drop to the liferaft is greater than 8 ft. There should be at least two exit routes from all areas of the survey vessel. Internationally designated symbols and colours should be displayed in all hazardous areas. This includes fire extinguisher types and areas that require hard hats, no smoking, wearing of lifelines, etc. Luminescent arrows should be placed at close intervals to indicate the route to the nearest exit.

Lifejackets should be stored in the cabins, one per person. A balance of lifejackets should be held on the upper deck, usually in a clearly marked locker and preferably in sufficient numbers for the entire crew. Two lifejackets should be carried for every person onboard. Lifejacket-donning instructions should be displayed in the cabins.

Survival suits should be carried that meet SOLAS requirements for immersion suits. These may be stowed in the cabins or at a stowage point near the muster point and displayed at all abandon ship drills. Survival suits are not a mandatory requirement in tropical waters. Lifebuoys should be carried with lights for a night-time man-overboard (MOB) incident. Typically, lifebuoys are fitted on the bridge wing extremities, in the well of the vessel and at the rear of the streamer deck. All buoys should be marked with the port of registry and the vessel's name. Ideally an MOB float should be fitted to the stern. An MOB raft may also be fitted to the stern. Inflatable rescue craft are considered suitable for MOB recovery.

An emergency position indicating radio beacon (EPIRB) should be carried operating on the world-wide emergency frequencies of 121.5 MHz and 243.0 MHz. EPIRBs should be stored in an accessible location and pointed out during safety initiation tours. The global maritime distress system (GMDSS) has now replaced all older radio methods of reporting distress.

6.9 Training and emergency response procedures

If inflatable rescue craft are carried the deployment and recovery method should be described in the safety program manual. At least two persons onboard should be specifically trained in respect of safety and rescue craft. Drills involving inflatable rescue craft should be carried out at least once a month. If inflatable rescue craft are used there should be specified weather limitations. There should be a written policy in respect of boat to boat personnel transfers. All personnel must wear personal flotation devices (PFDs) when in a dinghy or inflatable type craft. Helicopter transfers should be subject to rigorous safety routines but it is pointed out that very few site survey operations are equipped for helicopter landing; nevertheless, section 6.12 covers helicopter operations in some detail.

Box 6.2
Small boat accidents
Most survey operations use small boats from time to time. Mail, spares and other small pieces of equipment may need to be transported to or from a rig or rig supply vessel to the survey vessel. Small boats such as Zodiacs can be used for this type of transfer and may be used to change streamer depth control birds or perhaps carry out repairs to a streamer tailbuoy. Small boat work is always popular with the younger members of a survey crew. The trouble is that there have been a number of serious accidents involving small boats and some of these have resulted in fatalities. Small boat operations should be subjected to the most rigorous supervision. The survey vessel captain, survey party manager and oil company representative onboard should be informed before any small boat deployment and each should have the power of veto on small boat operations. The small boat crew should always include someone with a boatswain's safety certificate. The boat crew should wear proper helicopter type survival suits with watertight seals at the wrists and neck. Small boats should be used for essential transfers only and the number of deployments kept to an absolute minimum.

Fire drills should be held at least on a monthly basis but preferably bi-weekly and include a safety meeting of some 15–20 min duration. A fire drill should be held within 24 h of the vessel's departure from port. A ship's muster list should be prominently displayed and in muster drills all personnel should be accounted for in less than 10 min. The various emergency equipment should be demonstrated, such as BA sets, stretchers, fire suits, hydrants and hoses and fixed deluge systems. The audibility of all alarms should be routinely checked. Man-overboard, fire and abandonment drills should all constitute part of the normal ship's routine. A record of all drills should be formally recorded in the ship's log.

A man-overboard (MOB) routine should be in place onboard every survey vessel. An MOB liferaft is always a useful addition to the back deck of a survey vessel. There should be a dedicated MOB alarm with an alarm button on the back deck. MOB drills should be held at least once a month and a dummy should be available for the drills. The time taken to launch an MOB boat should be regularly clocked, and the time taken to recover the MOB dummy.

Above 75 per cent of seismic and ship's personnel should have attended a 1-week offshore survival and fire-fighting course. The oil company or other contracting authority should be informed of all personnel who have not performed the offshore survival and fire-fighting course. Party managers should have attended a geophysical field safety management course or equivalent. All seismic and ship's personnel should have undergone a seaman's medical examination prior to assignment to the vessel. A safety tour should be mandatory for all personnel joining seismic vessels for the first time. This should include the vessel layout and operation of safety equipment, first aid kits, alarm points and muster stations. An explanation of emergency procedures and safety regulations should also be provided.

The emergency response procedures should include an emergency response plan and a medical evacuation plan. These plans should be specific to each operating area and the operating conditions expected in a particular area. Methods, routes and contingency plans for various weather conditions should also be taken into account. In the case of a medical evacuation, medical information and travel documents should accompany the patient. These response and medical evacuation plans should be a part of every survey. Typical examples are given in Figures 6.4, 6.5 and 6.6.[9]

6.10 Seismic operations

All survey operations should be covered by safety procedures specific to a particular survey operation. These should cover launching and recovery of all equipment. The remarks made here cover a general approach to seismic operations.

In the case of airguns operating at 2000 psi safety procedures must be stringent. Uncovered hands or flesh should never be put in front of a jet of air. High pressure air can tear skin, penetrate blood vessels and force air, dust and even oil particles into the skin. Air penetrating blood vessels can cause embolism and death. Only qualified persons should operate airguns, gun handling equipment and air compressors. All other personnel should stay clear of equipment, including lines, rigging and booms while deploying, recovering and working on gun systems. Eye and ear protection should be worn if there is any prospect of high pressure air release and an eyewash station should be located in the gun area. The storage tanks, pipes, lines and fittings used to carry high pressure air must not be interfered with while the air is at 2000 psi in the system. Pressure-relieving valves and other safety devices should never be removed or modified except for repair by qualified personnel. Block valves upstream or downstream from a relief valve should be locked in the open position. When opening valves, the valve should be closed by half a turn after the maximum open position has been reached.

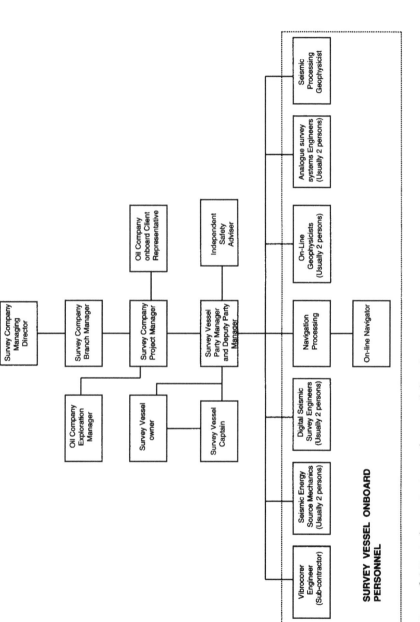

Figure 6.4 Typical organisation and responsibility chart.

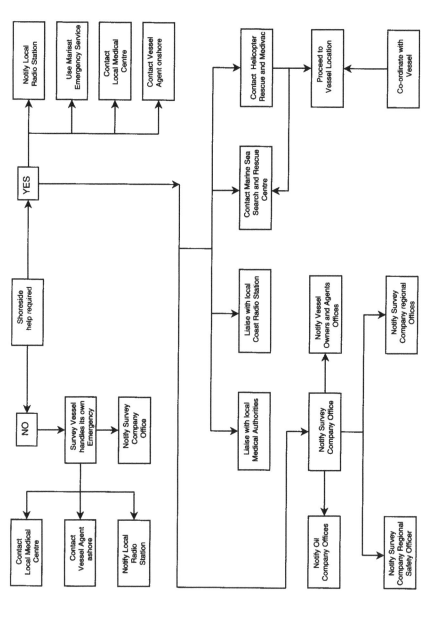

Figure 6.5 Typical emergency response plan.

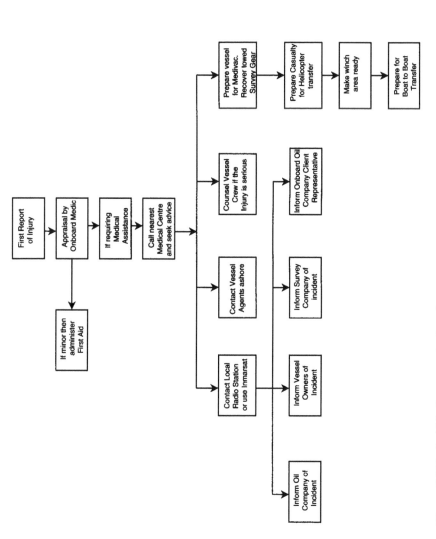

Figure 6.6 Typical Medivac organisation chart.

Compressed air can explode due to the high temperatures caused by compression (compression ignition) and by the addition of oil to hot compressed air. To minimise the risk of this happening valves should be opened slowly and the amount of oil in the system should be kept to a minimum. Air pressure should be bled off before recovery of seismic airguns. A red flashing light should be fitted to the gun-deck to warn personnel that airgun handling is in operation or that the guns on deck are under pressure. Airguns under test on deck should be worked at pressures of under 500 psi. Personnel exposed to water during gun recovery and retrieval should wear a safety harness and a personal flotation device. Personal safety equipment should also be worn. Every high pressure air injection injury, however small, should be treated as serious and reported.[10]

Box 6.3
The nastiest kind of accident
Airguns with air pressure in them are inherently dangerous. Airguns should never be pressured to 2000 psi on the deck of a survey ship. Even at 500 psi they are very dangerous and can inflict appalling injuries if they accidentally go off on deck. One such incident involved a survey operation with a large array of airguns. An inexperienced gun mechanic was running his hands over the surface of the airgun, looking for air leaks by feel. The gun autofired and blew his thumb off. It proved impossible to reattach the thumb. Looking for air leaks by feel is terribly dangerous. The correct method is to use an old washing up liquid bottle filled with soapy water. Spray the gun with the soapy water and look for soap bubbles caused by the air leak. Such extemporised methods are crude, simple and – much more to the point – safe.

Streamer deployment and recovery should be based on a clearly understood written procedure. The bridge must be informed before any streamer deployment or recovery is made and close co-operation maintained between bridge and back-deck. Expected weather conditions should be checked before deployment and the officer of the watch must monitor all vessel traffic in the area and warn the back-deck personnel of any course changes. CCTV equipment should monitor the back-deck operations and these should be observed on the bridge and in the instrument room. The streamer reel, brakes, hydraulic power, tailbuoy lights, batteries, radar reflectors, etc. should all be tested before streamer deployment. Safety harnesses and anchor points must be provided to the rear of the streamer deck. Safety harnesses and flotation devices should be worn as for gun deployment and recovery. Streamer reels should be guarded from gangways and access ways by railings and only operated by experienced persons. Lifting hoists for streamer tailbuoys should be suitable for the expected weight of the load plus the expected shock load caused by vessel movement in rough seas. No-one should be between the tailbuoy and the vessel stern during tailbuoy deployment. Care should be taken when stopping the streamer reel to attach or remove the depth control birds. Streamer reels should be secured when deployment is complete. Reel brakes can 'creep' when unattended. Streamer oil is highly flammable and oil should be immediately cleared away after spillage. Streamer oil drums must be clearly marked. No smoking, welding or open flames should be allowed anywhere near the streamer reel.

Box 6.4
Safety issues for streamer deployment and recovery
Streamer reels and the operating area around them are the source of numerous minor accidents. Many older survey vessels have wooden-planked streamer decks which tend to become oil soaked from the streamer kerosene and a fire hazard as a result. The streamer deck should be degreased regularly and sprinkled with sand before, during and after streamer handling. Remarks about degreasing also apply to the gun handling, maintenance and repair areas. Streamer handling and repair involves much use of Stanley knives, to strip off electrical insulation tape, etc. The most common form of accident is a deep cut at the base of the thumb or other part of the hand, inflicted with a Stanley knife that slipped. Heavy gloves, even ordinary gardening gloves, are invaluable protection. Enclosed Stanley blade insulation strippers are also available and should be used whenever possible.

Battery-charging compartments should be well ventilated to avoid gas accumulation. All light fittings in battery compartments should be fitted with protective glass. Smoking in battery compartments is absolutely prohibited. Storage of other equipment in battery compartments is absolutely prohibited. Batteries should be battened into position. Alkaline and lead–acid batteries should be in separate compartments. If spattering or injury is caused by lead–acid cells then liberal application of water will dilute the sulphuric acid. The same spattering from alkaline cells may require neutralisation with a boracic powder saturated solution. Lithium batteries, widely used in streamer depth control birds, are potentially very hazardous and can explode if shorted or heated. Disposal of lithium batteries should always be according to manufacturer's instructions.

6.11 Electrical equipment and emergency power supplies

Adequate lighting should be provided where crew members are working and particular attention should be paid to stairs and ladders. Portable lights should never be lowered by their leads. Leads should be kept clear of running gear,

210 *High resolution site surveys*

moving parts and machinery. If leads pass through doorways they should be latched open. Every installation and circuit should be protected against overload by automatic devices. All electrical equipment cabinets and housings should be grounded. No fuse or circuit breaker other than a linked circuit breaker should be inserted in a grounded neutral conductor. Any linked circuit breaker inserted in a grounded neutral conductor should be arranged so as to break all related energised conductors. A single pole switch should be inserted in the energised conductor only. When maintenance is required on electrical power lines, motors, equipment or fuel-powered engines a lockout/tagout system should be in operation.

The number and type of the ship's electrical generators should be checked before the start of survey. As a basic rule any survey vessel should have two generators, one of which is a standby generator, capable of supplying the vessel with its electrical power requirements in the event of the primary generator failure. Emergency battery power supplies should be carried and should provide sufficient power for essential bridge equipment. This would include the Marisat communications system, the ship's radar and autopilot and any bridge equipment necessary for the safe operation of the vessel. Battery power should also provide emergency lighting throughout the ship, to allow a safe evacuation of the personnel from the ship's interior, if this became necessary.

6.12 Helicopter operations

Before survey a description of the ship's helideck surface, dimensions and any obstructions should be supplied to the helicopter company. The obstacle free angle of the helideck should be known and any refuelling facilities in the area. For a particular survey area alternative landing sites such as rigs, platforms, islands, etc. should be identified. Equally, the type, weight and rotor length of the helicopters to be used for a particular survey should be known by the survey company.[11]

A landing net should be available for use at all times. This net should be secured at 1.5 m intervals around the perimeter of the landing area so that it is stretched taught. The net must be checked by the helicopter landing officer (HLO). All personnel must exercise care when approaching the helicopter over the net. Landing lights and floodlights should be checked before every helicopter operation. Fire-fighting equipment, particularly in regard to adequate chemical foam, should be checked prior to each landing. The helicopter fire-fighting unit should be staffed and ready at every helicopter take-off and landing. The firefighter should wear a total fire protection suit made from aluminium. The ship's aqueous film-forming foam (AFFF) system should be able to provide foam to the stern of the landing area from two hoses each on its own branch line. One or more dry powder fire extinguishers totalling 45 kg and one or more CO_2 or Halon 1211 extinguishers totalling 18 kg should be available near the helicopter landing area. The minimum crew required for helicopter operations should include an HLO, a trained firefighter in a protective suit with breathing apparatus, a baggage handler and a fire valve attendant. When the helicopter lands, the HLO and firefighter should have a clear view of the helideck. The helideck should be clear of obstructions and all radio-communication systems should be checked. An emergency equipment box should be nearby containing axes, bolt cropper, heavy duty hacksaw, seat-belt cutting knives etc. It should also contain a self-contained breathing apparatus, a safety harness, a fireproof lifeline and a battery-operated, hand-held lantern.

Before an actual flight the operating radio frequency and call sign of the helicopter and the estimated time of departure and arrival must be known. Contact with the helicopter should be made as soon as possible after the estimated time of departure (ETD) of the helicopter. The helicopter must be furnished with a weather report from the survey vessel, the vessel position, course, speed and heading. Details of the survey vessel non-directional radio beacon (NDB), including the NDB frequency and code, should also be supplied and the time this beacon will be switched on. When the helicopter is within visual range and a positive identification of the survey vessel has been made, the change of radio frequency for landing clearance should be made. The number of passengers, weight of baggage and weight of any additional cargo should be communicated to the helicopter operator. In the case of a medical evacuation the extent of injuries and the number of stretcher cases should be communicated. The ship's crane and work boat should be stowed with all protective covers lashed securely. The rescue boat should be ready for launch during all helicopter operations.

During helicopter operations no-one should approach the helicopter when its red flashing anti-collision light is on. The tail and air intakes should be avoided at all times. An approach to the helicopter should be made only within line of sight of the helicopter crew. When in the helicopter seat belts, ear protectors and flotation devices should be worn. Passengers should familiarise themselves with the emergency exits and emergency equipment locations. Upon landing, passengers should remain seated until instructed to disembark by the flight crew.

Onboard the survey vessel a cargo handler should be briefed on the type of helicopter in use. This person will open and close the cargo doors and be familiar with the cargo lashing points. If fuelling is to take place all fuel pumps, motors, hoses, nozzles and fuel pump filters should be of an approved type. All nozzles and fittings should be manufactured from non-sparking material. All electrical systems should be grounded. Fuel storage should be at least 15 m from any power source and there should be no smoking within 15 m of the fuel source. An approved fuel filtering system with water and contamination separation must be used in conjunction with the fuel storage and refuelling facilities.

6.13 Personnel competence assurance

A modern approach to personnel safety may be defined under the general heading of competence assurance. The International Marine Contractors

212 *High resolution site surveys*

Association (IMCA) have recently developed a competence assurance and assessment scheme. This approach to safety assumes that every individual associated with a survey operation in some way makes an impact on the total safety environment. Some jobs are obviously more safety critical than others. Certain 'core competencies' should be considered essential to all personnel working in the survey industry. Essentially, these are defined by the requirement that all offshore personnel should have completed a 3-day safety and survival course before employment offshore. Each safety critical task or operation carries with it 'key competencies'. These competencies define specified criteria that determine whether an individual has the required knowledge and ability to perform a certain safety critical task. It is probable that in the not too distant future, individuals working offshore may carry an IMCA Record of Competence Log Book covering personal details, qualification certificates, training, specialist experience and details of all competence assessments.[12]

This means that all personnel working offshore in safety critical positions will be allocated a job function within a framework of minimum entry qualifications. As people broaden their skill base and their experience accumulates, their improving competence will be reviewed and assessed against specified criteria with progress recorded in the IMCA Record of Competence Log Book. As part of this scheme the oil companies and survey contracting companies should have internal auditing arrangements agreed with the IMCA. A range of criteria will be involved in these assessments, including academic and vocational qualifications, demonstrable experience, technical skills, medical certification and appropriate training.

Certain core competencies are essential for people working offshore. These include safety issues, behavioural factors, emergency response and communication. To core competence must be added key competence, where assessors apply specified criteria which will determine whether a person has the required knowledge and can demonstrate their ability to complete a particular task.

Assessors can be any supervisor or manager with the necessary background knowledge to apply the IMCA criteria in a fairly rigorous manner. Every IMCA company will identify its own assessors and train them where required.

6.14 Environmental monitoring

In recent years there has been increasing concern for the environment and the activities of the seismic industry have come under the scrutiny of groups such as Greenpeace. As part of the United Kingdom's response to the Agreement on the Conservation of Small Cetaceans in the Baltic and North Sea (ASCOBANS), in February 1995 the then Department of the Environment issued guidelines for minimising acoustic disturbance to small cetaceans. This came about because the airguns used for seismic surveys generate sound that is mostly low frequency, overlapping with the frequencies produced by baleen whales. These mammals are considered to be vulnerable to disturbance from seismic surveys. Toothed whales and dolphins use higher frequencies for communication and echolocation.

The guidelines were revised by the Joint Nature Conservation Committee (JNCC) in early 1996. Under this version of the guidelines, operators are required to consult the Joint Nature Conservation Committee when planning seismic surveys in the UK and, if necessary, discuss precautions which can be taken to reduce disturbance. During survey, before the start of line, operators are required to check for the presence of cetaceans and delay the start of line by at least 20 min if cetaceans are within 500 m. Whenever possible, a soft start procedure should be employed, building the airgun array power over 20 min from a low starting energy. How far this applies to site surveys, where the energy source is often only 2–5 per cent of the power output of a conventional seismic source, is at present unclear.[13]

It is likely to be the case that all survey operations will be asked to carry out cetacean watches during survey.

References

1. E and P Forum Schedules in the possession of the author. At least one site survey contract stated these Forum Schedules.
2. *UKOOA Guidelines for the Conduct of Mobile Drilling Rig Site Surveys*, Volume 1, pp. 2 and 3.
3. Taken from a safety audit performed to oil company specifications.
4. Taken from a vessel safety audit performed by the author.
5. Taken from a site survey contractor's safety case on a particular vessel.
6. Taken from a safety audit report written by the author.
7. Marine Geophysical Operations safety manual.
8. Taken from a safety audit written by the author.
9. Taken from various contractors' safety cases for particular operations.
10. Marine Geophysical Operations Safety Manual.
11. Marine Geophysical Operations Safety Manual and Warsash offshore safety course, 'Further Offshore Emergency Training'.
12. 'Competence assurance and assessment scheme', *IMCA Newsletter*, April 1999.
13. *Cetacean Observations During Seismic Surveys in 1997*, JNCC Report number 278.

Other sources consulted

Warsash safety course, First Aid at Sea course notes.
GSR (Geophysical Safety Resources) Safety Management Course Notes.
Geocon Group of Companies, Health and Safety Protection of the Environment Policy Statement.
Geocon safety audit papers, Outline and Approach.
Two emergency response plans have been quoted, one from an oil company and one from a survey company.
Five safety audits were used in the preparation of this chapter.
Geocon IAGC Consultants Committee Code of Practice.

Glossary and notation

A/D   Analogue to digital. The conversion of analogue signals into a digital form.
AFFF   Aqueous film-forming foam.
AGC   Automatic gain control.
Airgun   A marine seismic energy source which injects a bubble of highly compressed air into the water. Oscillations of the bubble as it alternately expands and contracts generate a sound wave whose frequency depends on the amount of air in the bubble, its pressure and the water depth.
Alias   An ambiguity in the frequency represented by sampled data. Where there are fewer than two samples per cycle, an input signal at one frequency yields the same sample values as another frequency. Half of the frequency of sampling is called the Nyquist frequency. If a signal is sampled every 4 ms, or 250 times a second, the Nyquist frequency is 125 Hz. If 50 Hz is within the passband, then 200 Hz will also be passed if an alias filter has not been used.
Alias filter   A filter used to remove undesired frequencies which the digital sampling process would otherwise alias.
Anomaly   Deviation from uniformity on a seismic record.
Anti-alias   Alias filter for digital recording systems. See alias filter above.
Anticline   A fold or arch away from which the strata dip in opposite directions.
Aperture   Gate or window.
Argo   Older type of radio-navigation system. Obsolete.
AVO   Amplitude versus offset.
Azimuth   The horizontal angle, usually clockwise, from true north.
Bandpass   The frequency range passed by a system with minimal attenuation. Frequencies outside the bandpass limit are massively attenuated.
Bandwidth   The range of frequencies over which a system is designed to operate.
Bar   Unit of acoustic pressure, 10^5 N/m^2.
Baseline   A reference line usually used for calibrating positioning systems.
Base station   A reference station used for radio positioning systems.
Baud   The ability of a channel to transmit information. One bit per second.

Binary gain A gain control system in which the gain is changed in discrete steps by a factor of two.
Bird Depth control unit attached to a streamer.
Bird dog The oil company representative on a seismic field crew (American). Slang term in widespread use.
Bit A binary digit, the smallest unit of information.
Block diagram Diagram showing the functions of a system and how they are interrelated without showing construction details.
Boomer A marine seismic energy source in which capacitors are charged to a high voltage and then discharged through a transducer in the water.
BOP Blow-out preventer.
Bright spots Amplitude anomalies associated with high pressure gas pockets.
Cable An assembly of hydrophones towed behind a survey ship. Also known as a streamer.
Caesium Caesium magnetometer, a type of optically pumped magnetometer.
Camera A recording oscillograph usually producing records on photosensitive paper or film.
Capacitance The ratio of charge (Q in coulombs) on a capacitor to the potential across it (V in volts) is the capacitance (C in farads).
Cavitation The situation where the pressure in a liquid becomes lower than the hydrostatic pressure. The collapse of the liquid into the region generates a shock wave by implosion.
CDP Common depth point (method).
CGG Compagnie Général de Géophysique. A geophysical contractor.
Convolution Change in wave shape as a result of passing through a linear filter, a mathematical operation between two functions, or the act of linear filtering.
Core A rock sample taken from a borehole and a sea bed sample taken as a long thin circular section.
Coriolis effect Acceleration, in the rotating co-ordinate system of the Earth, of a body in motion with respect to the Earth resulting from the rotation of the Earth.
CPT Cone penetration test.
CRT Cathode ray oscilloscope.
CW Continuous wave.
D/A Conversion of a digital, usually binary number, into a corresponding voltage.
DAS Deconvolution after stack.
DBS Deconvolution before stack.
Decca A medium range positioning system, using the phase comparison of synchronised CW radio waves transmitted from three or more radio stations. Obsolete.

Decibel (dB) A unit used in expressing power or intensity ratios: $20 \log_{10}$ of the amplitude ratio, or $10 \log_{10}$ of the power ratio. An amplitude ratio of 2 represents a power ratio of 4 and is equivalent to 6 dB.
dB/octave Unit for expressing the slopes of filter curves.
Deconvolution The process of undoing the effect of another filter.
Demultiplex Separating the individual component channels which have been multiplexed.
Depth controller A device with movable vanes that maintains a seismic streamer at a particular depth (see bird).
DFS Digital field system.
DGPS Differential global positioning system.
Diapir A flow structure whose mobile core has pierced overlying rocks.
Diffraction Scattered seismic energy which emanates from an abrupt discontinuity of strata type.
Digital Representation of quantities in discrete quantised units.
Dip The angle made by a plane surface with the horizontal.
Doghouse Enclosed space containing the seismic instrumentation. Slang term of American origin in widespread use.
Dogleg An abrupt angular change in direction of, say, a survey line.
Doodlebugger A geophysicist working on a field crew. Derogatory term of American origin in widespread use.
DOP Dilution of position.
Doppler effect Apparent change in frequency of a wave as the source moves towards and then away from an observer.
DOS Disc-operated system.
DTI Department of Trade and Industry.
E and P Exploration and production.
Easting Easterly component of a survey computation.
Eccentricity The ratio of the focus-to-centre distance to the length of the semi-major axis for an ellipse.
Echosounder A device for measuring water depth by timing sonic reflections from the water bottom.
EG and G Edgerton, Germhausen and Grier (equipment manufacturers).
Ellipsoid A solid figure for which every plane cross-section is an ellipse.
Eotvos effect The vertical component of the Coriolis acceleration observed while measuring gravity in motion.
Ephemeris Message broadcast from a GPS satellite to inform the user of health and position of satellite.
EPIRB Emergency position indicating radio beacon.
Fathometer Alternative name for an echosounder.
Feathering The effect of cross-currents on a towed hydrophone array. The angle made by the array (streamer) to the plotted survey line is the feather angle.
FET Field effect transistor.

FFT Fast Fourier transform.
Fiducials Time marks on a seismic record, as in fiducial marks.
First break The first recorded signal attributed to a seismic impulse from a known source.
Fix A determination of location on a survey record.
F–K A domain in which the variables are frequency and wave number.
Flattening The ratio of the difference between the major and minor axes of an ellipse to the major axis.
Flexichoc An implosive energy source (now obsolete).
Floating point A number expressed by a certain number of significant figures times a base raised to a power.
Fluxgate Fluxgate magnetometer.
FM Frequency modulation.
Fold Common depth point multiplicity.
Fourier The analytical representation of a waveform as a weighted series of sinusoidal functions.
Fourier transform The formula which converts the time function into the frequency domain representation.
***F*-test** A statistical test for DGPS. It is a measure of the overall fit of the observations to the model.
Gal A unit of acceleration used in gravity measurement. $1\,\text{gal} = 10^{-2}\,\text{m/s}^2$. The Earth's nominal gravity is 980 gals.
Gamma Unit of magnetic field. $1\,\text{gamma} = 10^{-5}\,\text{G} = 10^{-9}\,\text{T}$. The gamma has now been replaced by the nanotesla.
Geodesy The location of points on the Earth's surface with respect to reference systems.
Geodetic latitude Ordinary latitude. The angle between a tangent to the ellipsoid which approximates the Earth's shape and the Earth's axis.
Geoid The sea-level equipotential surface to which the direction of gravity is everywhere perpendicular.
GLONASS Global'naya Navigatsionnaya Sputnikova Sistema. The Russian equivalent of GPS.
GPS Global positioning system.
Gravimeter An instrument for measuring variations in gravitational attraction.
Greenwich Longitude measured with respect to the prime meridian which passes through the Royal Astronomical Observatory at Greenwich.
Group interval The horizontal distance between the centres of adjacent hydrophone groups.
GSI Geophysical Service International. Now defunct geophysical contractor.
Gyrocompass A gimbal-mounted gyroscope incorporating unbalanced masses which makes the axis of rotation precess about true north.
Harmonic A frequency which is a simple multiple of a fundamental frequency.
HDOP Horizontal dilution of position.

Header The identification information and tabulation data which precedes seismic data on a data tape.
Hertz Unit of frequency (Hz).
HF High frequency.
Hi-fix An older Decca positioning system. No longer in use.
High cut filter A filter that transmits frequencies only below a given cut-off frequency.
Hiran High accuracy Shoran. Obsolete.
HLO Helicopter Landing Officer.
Hooke's law Hooke's law states that stress is proportional to strain.
HSE Health and Safety Executive.
Hyperbolic A hyperbolic line of position determined by measuring the difference in distance to two fixed points.
Hyperfix Older type of hyperbolic positioning system. Now obsolete.
IAGC International Association of Geophysical Contractors.
IFP Instantaneous floating point.
IMCA International Marine Contractors Association.
IMO International Marine Organisation.
Impedance The resistance to the flow of alternating current.
Inertial navigation A dead-reckoning method of determining position in which accelerations are measured with very sensitive accelerometers and doubly integrated to give position.
Ionosphere The layers of atmosphere between 50 and 1000 km above the Earth.
Kalman filter A recursive filtering system widely used in the seismic industry.
Kirchhoff Kirchhoff's first law states that the vector sum of currents at a junction is zero. The second law states that the vector sum of all currents around a loop is zero.
Knot A nautical mile per hour (1.1508 statute miles, 1.852 km/h).
Lambda Medium range hyperbolic positioning system. Obsolete.
Lane The distance represented by one cycle of the standing wave interference pattern resulting from two radiated waves. Used in the context of hyperbolic positioning systems.
Larmor frequency The frequency with which gyromagnetic moments precess in a magnetic field. Term associated with magnetometers.
LAT Least astronomical tide.
Latitude The distance north or south of an east–west reference line.
Leakage Low electrical resistance to ground where there should be high resistance. Term associated with seismic streamers.
LF Low frequency.
LOP Line of position.
Loran Long range navigation. Hyperbolic positioning system. Obsolete.
Low cut filter A filter that transmits frequencies above a given cut-off frequency and substantially attenuates lower frequencies.

LRTK Long range real time kinematic.
Magnetometer An instrument for measuring magnetic field strength.
Magnetostriction Change in the strain of a magnetic material as a result of changes in magnetisation.
Maxiran Older type of ranging navigation system. Obsolete.
MDE Marginally detectable error.
Mercator projection A conformal cylindrical map projection developed on a cylinder tangent along the equator with the expansion of the meridians equal to that of the parallels.
MF Medium frequency.
Microfix Decca ranging navigation system. Obsolete.
Migration The plotting of dipping reflections in their true spatial positions.
Milligal Unit of gravitational measurement, 10^{-5} m/s^2.
Mini-sleeve exploder A seismic sound source in which a gas (propane or butane) is exploded to produce the pulse while the venting of gases into the atmosphere rather than the sea reduces the bubble oscillation.
Moveout The difference in arrival time at different hydrophone positions.
MSL Mean sea level.
Multiples Seismic energy which has been reflected more than once.
Multiplexed format A sequence of data in which the first sample of channel 1 is followed by the first sample of channel 2, etc.
Near field The field near a source.
Near trace gather A record section comprising data from the hydrophone group nearest the shot.
NMO Normal moveout. The variation of reflection arrival times because of variation in the shotpoint to hydrophone distance (offset).
NSRF Nova Scotia Research Foundation.
Nuclear Nuclear precession magnetometers utilise nuclear resonance. The resonant frequency is proportional to the absolute magnetic field strength.
Nyquist frequency A frequency associated with sampling which is equal to half the sampling frequency. Also called the folding frequency. Frequencies greater than the Nyquist will alias as lower frequencies from which they are indistinguishable.
Octave Separation of two frequencies having a ratio of 2 or $\frac{1}{2}$. Filter roll-off is usually given in dB/octave.
Offset The distance from the centre of the sound source to the centre of the near hydrophone group.
Optically pumped Magnetometers using caesium or rubidium vapour are optically pumped. Atoms in a vapour cell precess about the steady magnetic field which is to be measured.
Passband The range of frequencies which can pass through a filter without significant attenuation.
PCPT Piezocone penetrometer test.
PDOP Precision dilution of position.

PFD Personal flotation device.

Piezoelectric The property of a dielectric which generates a voltage across it in response to a mechanical stress.

Pinger A shallow penetration high power high frequency seismic profiling system used to delineate shallow strata immediately below the seabed.

Precession The tendency of a gyroscope to turn under the influence of a torque which tries to change the direction of its axis of spin.

Pseudo-range The time shift required to line up a replica of the code generated in the received code from the satellite × the speed of light.

Pulse-8 Decca time comparison positioning system. Obsolete.

P-wave An elastic body in which particle motion is in the direction of propagation. The type of seismic wave produced by all marine seismic energy sources.

P2/91 UKOOA seismic navigation raw navigation data format.

P1/90 UKOOA seismic navigation post-plot navigation data format.

QA Quality assurance.

QC Quality control.

Quaternary gain A gain control system in which amplitude is changed in discrete steps by factors of 4. Compared with binary gain, fewer gain jumps are required for a recording and an additional bit is required in the gain word, but one bit is saved in the amplitude recording to still have the same range of gain.

Rana Obsolete medium frequency positioning system.

Raydist Medium frequency, medium range hyperbolic positioning system. No longer in use.

Rayleigh–Willis The oscillation period T for the bubble effect varies as the cube root of the energy Q (in joules) and inversely as the 5/6 power of the pressure. $T = 0.0450 Q^{1/3}/(D+33)^{5/6}$.

RDU Remote data unit.

Reflection The energy or wave from a shot or other seismic source which has been reflected from an acoustic-impedance contrast (reflector).

Refraction The change in direction of a seismic ray in passing into a medium of different velocity.

Resolution The ability to separate two features close together.

Reverberation Multiple reflection in a seismic layer.

RIDDOR Reporting of injuries, diseases and dangerous occurrences.

RMS Root mean square.

Roll-off The frequency beyond which a filter produces significant attenuation. Usually stated in dB/octave.

RTCM-SC104 US Radio Technical Commission for Maritime Services Special Committee 104. Set to define a format for the GPS correction messages from reference stations.

RTK Real time kinematic.

Satnav Satellite navigation.

SBL Short baseline.

SCSI Small computer system interface.
SEG Society of Exploration Geophysicists. Digital tape formats.
Sensitivity The least change in a quantity which a detector can detect. An instrument can have excellent sensitivity but poor accuracy.
Shoran Short range navigation system where the distances from reference stations are determined by measuring the travel time of pulsed radio waves. This system became obsolete in the late 1970s.
SI Système International (units).
Sidescan sonar A method of locating irregularities on the sea floor. A pulse of sonar energy at typically 120 kHz is emitted from a towed unit known as a towfish. The sonar beam is narrow in the direction of traverse because the source consists of a linear array of elements. Bottom irregularities and variations in bottom sediments produce changes in the energy return.
Siemens Unit of electrical conductivity. The reciprocal of ohms. Also called mho.
Sigma–delta A digital recording system where each channel has its own A/D.
Signal-to-noise ratio The energy of desired events (signals) divided by all remaining energy (noise) at that time.
Signature Waveshape that is characteristic of a particular source.
Sky-wave Electromagnetic radio waves which reflect from ionised layers in the ionosphere.
Sleeve exploder A marine seismic energy source in which a gas is exploded in a thick rubber sleeve from which the waste gases are vented into the atmosphere rather than into the water.
Smoothing Averaging adjacent values.
SOLAS Safety of life at sea.
Sonobuoy Device used in refraction surveys for detecting energy from a distant shot and radioing the information to the survey ship.
Sparker A seismic energy source in which an electrical discharge in water is the energy source.
Spheroid The oblate ellipsoid of revolution used to approximate the Earth's shape.
SSBL Super short baseline.
Stack A composite record made by mixing traces from different records.
Streamer A marine cable, designed for towing by a ship, incorporating pressure hydrophones. A marine streamer will have typically twenty-four or forty-eight 'live' or 'active' sections with perhaps eighteen hydrophones per 'live' or 'active' section.
S-wave Shear or transverse wave. Marine seismic energy sources do not produce shear waves because water has no shear characteristic.
Syledis 430 MHz ranging navigation system. Obsolete.
TAR True amplitude recovery.
Telemetry The transmission of data from a point of observation to a point of recording, often by radio.

TEMPSC   Totally enclosed motor-propelled survival craft.
Tesla   Unit of magnetic field strength (T). $1\,T = 1\,Wb/m^2 = 10^4\,G$.
TI   Texas Instruments.
Time break   The mark on a seismic record which indicates the shot instant or time at which the seismic wave was generated.
Toran   Medium range phase comparison positioning system. Obsolete.
Trace   A record of one seismic channel.
Trace sequential   A format on magnetic tape in which one channel is recorded without interruption followed sequentially by other channels.
Transducer   A device which converts one form of energy to another.
Triangulation   Establishing locations by a system of overlapping triangles. The angles are directly measured and the intersection triangle defines the (survey ship) position.
Trisponder   An older type of radar-based positioning system that used triangulation for defining position.
Tropospheric scatter   The bending of radio waves in the atmosphere by scattering instead of refraction.
Turkey shoot   Direct comparison of two or more seismic systems under the same conditions over the same area.
TVF   Time variant filter.
TVG   Time variant gain.
TWWT   Two-way travel time.
UK   United Kingdom.
UKCS   United Kingdom continental shelf.
UKOOA   United Kingdom Offshore Operators Association.
UTM   Universal transverse Mercator. A standard rectangular grid map. The projection is onto a cylinder tangent to the Earth along a central meridian.
VDOP   Vertical dilution of precision.
VHF   Very high frequency.
VLF   Very low frequency.
Waterbreak   The arrival of energy which travels through the water directly from the seismic sound source to the waterbreak detector.
Water gun   A seismic sound source that produces its acoustic output from the sudden collapse of a cavitation volume in the water.
Wavelet   A seismic pulse usually consisting of $1\frac{1}{2}$ –2 cycles.
WGS-84   World Geodetic System 1984.
WHO   World Health Organisation.
w-test   The w-test is used to prove the null hypothesis by testing it against a series of alternative hypotheses.
Yaw   Rotational motion of a ship about a vertical axis.
2D   Two dimensional.
3D   Three dimensional.

Glossary 223

Base units

Length	Metre (m)
Mass	Kilogram (kg)
Time	Second (s)
Current	Ampere (A)
Temperature	Kelvin (K)
Plane angle	Radian (rad)

Multiplication factors

10^9	Giga (G) as in gigahertz
10^6	Mega (M) as in megahertz
10^3	Kilo (k) as in kilohertz
10^{-3}	Milli (m) as in millivolts
10^{-6}	Micro (μ) as in microvolts
10^{-9}	Nano (n) as in nanoseconds

Derived units

Volume	Cubic metre (m^3)
Force	Newton (N)
Energy, work	Joule (J)
Frequency	Hertz (Hz)
Acceleration	Metre per second squared (m/s^2)
Potential	Volt (V)
Capacitance	Farad (F)
Magnetic field strength	Tesla (T)
Area	Square metre (m^2)
Density	Kilogram per cubic metre (kg/m^3)
Pressure	Pascal (Pa)
Power	Watt (W)
Velocity	Metre per second (m/s)
Charge	Coulomb (C)
Resistance	Ohm (Ω)
Magnetic flux	Weber (Wb)
Inductance	Henry (H)

Greek alphabet

A	α	alpha	P-wave velocity, angle
B	β	beta	S-wave velocity, angle
Γ	γ	gamma	Phase or phase difference
Δ	δ	delta	Δ: sampling interval, dilation; δ: delay time or angle
E	ε	epsilon	Eccentricity, phase shift

224 *High resolution site surveys*

Z	ζ	zeta	Vector displacement
H	η	eta	Absorption coefficient
Θ	θ	theta	Angle
I	ι	iota	Incident angle
K	κ	kappa	
Λ	λ	lambda	Wavelength
M	μ	mu	micro
N	ν	nu	Frequency
Ξ	ξ	xi	ξ: dip; Ξ: strike
O	o	omicron	
Π	π	pi	3.142
P	ρ	rho	Density
Σ	σ	sigma	Standard deviation
T	τ	tau	Damping factor
Y	υ	upsilon	
Φ	φ	phi	Φ: transform of P-wave
X	χ	chi	S-wave displacement
Ψ	ψ	psi	Wavefunction
Ω	ω	omega	Ω: ohms; ω: angular frequency

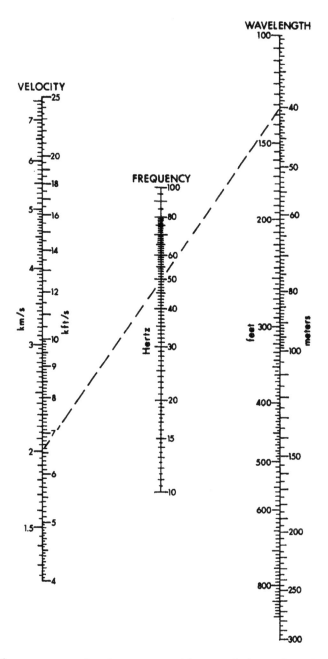

Velocity, frequency, wavelength nomogram. The straight line relates the velocity, frequency and wavelength. The scales shown are metric on the outer and imperial on the inner. As an example, a velocity of 2 km/s and a frequency of 50 Hz gives a wavelength of 40 m. (Taken from Sheriff and Geldhart, *Exploration Seismology*, vol. 1, p. 43 and reproduced by kind permission of the Cambridge University Press.)

Index

A/D, 34, 37, 39,
ADC, 40
AGC, 31, 37, 39, 55, 70, 72, 131
Aimers Mclean corer, 136
airgun, 7, 44, 51, 52, 56, 57, 60
ALBAC, 108
Alpine Geophysical, 44
amplitude anomaly, 69, 75, 76
amplitude versus offset (AVO), 67, 72, 75, 76
analogue data, 8
analogue module, 25, 26, 28, 29, 30, 31
analogue to digital, 7, 31, 32
anchor, 9, 13, 17
Anderra current meter, 138
angles of incidence, 2
anti-alias filter, 31, 32, 37, 41, 42
Aqua Explorer 2, 109
Aquatronics *see* Fairfield-Aquatronics
Argo, 14
Atlas-Deso 20, 81, 82, 83
autonomous underwater vehicles (AUV)s, 10, 14, 81, 97, 104, 105, 106, 107, 108, 186
Autosub, 107
axial thrusters, 19

bandwidth, 6, 37
bar check, 84
bathymetry, 10, 15, 21, 85, 95, 101, 104
beamwidth, 8
bedrock, 8
Benson-Schlumberger, 130
Benthos, 134
binary, 7, 24
birdog, 21
Bismarck, 14
blowout, 3, 5

boomers, 8, 9, 15, 81, 93, 115, 118, 119, 121, 123, 124, 130, 131
bright spots, 4, 16, 24, 73, 74, 77
brute stack, 37, 67

CA-code, 156, 158, 159, 165
caesium vapor magnetometers, 12, 140, 148, 149
Cape Hatteras, 14
catamaran, 9, 118
Cauer (filter), 31
CDP, 3, 73
Central Processing Unit (CPU), 34
CGG HR 6300, 24
Chilowsky, 2
chirp profiling, 111, 140
Client Representative, 21
comb sparker, 9
commercially available (CA) code, 14
common depth point, 3, 4
common mid-point (CMP), 65, 70, 72
common mode rejection, 37, 42, 43
controller module, 25, 28, 29, 30,
converter, 24, 38, 39
crossfeed, 38, 43
crosstalk, 37
current meters, 11, 137

DAC (Data Acquisition Controller), 34
DAS 1A *see* OYO and Geospace, 24, 28, 34, 35, 37, 40, 41, 42, 43
Dassault-Sercel Navigation Positioning (DSNP), 156, 173, 174
Datasonics SIS-1000, 111
DC offset, 40, 41, 42, 43
DCM (Display Controller Module), 58, 59
Decca Arkas, 18
deconvolution, 67, 69, 70, 71, 73

Index

deconvolution after stack (DAS), 71
deconvolution before stack (DBS), 70, 73
deep-tow, 9
Delph, 131
delta-sigma, 32
demultiple, 69
demultiplex, 25, 31, 34, 132
depth control, 61, 62, 65
designature, 69,
DFS III, 24,
DFS IV, 24
DFS V, 24, 25, 28, 29, 30, 37, 60, 61, 64
DGPS, 14, 91, 106, 166, 169, 170, 171, 173, 174, 175, 176, 177, 179, 181, 186, 190
diapirs, 70, 71
diesel-electric, 19
Digicourse, 62, 65, 66
digital computers, 7
digital to analog, 31
digital filtering, 34
digital recording systems, 7, 24, 34, 58, 67–68, 115, 131
digital seismic acquisition, 5,7
digital seismic data, 8
digital seismic processing, 8
dip moveout (DMO), 67, 69, 70
distortion, 37,
DMX *see* Geometrics Strativisor, 31
DRAM, 34
drop corers, 11
dynamic offset, 28
dynamic range, 24, 31, 32, 36, 37, 38, 41

echosounders, 8, 10, 13, 14, 15, 81, 82, 84, 85, 95, 101, 102, 105, 106, 109, 111, 112, 140, 185
EDO-Western, 111
EG and G, 47, 88, 93, 118, 121, 126
Elac Bottomchart, 98
electromagnetic, 8
environmental monitoring, 11
EPC graphic recorders, 130
exponential oscillator, 39
Exxon, 57

Fairfield-Aquatronics, 6, 24, 44, 57
farfield test, 53
fauna, 11
feathering, of streamers, 6, 7, 65, 66, 78
Ferranti 310C, 111, 118

Fessenden, 81
filter, 30, 31, 32, 34, 38, 39, 40, 43, 72, 104
fixed pitch propellers, 19
Fjord, 7, 60, 61
F-K filters, 69, 71
Flexichoc (Geomechanique), 57
floating point, 24, 25, 39
flora, 11
Focus ROTV, 140
Fourier transform, 40
Fresnel zones, 71
F-test, 172, 176
Fugro-Geoteam, 156
full-fold, 7, 28

Gambas, 131
GCCC digital streamer, 34, 61
Geco, 7, 61
Geoacoustics, 90, 101, 111
Geoconvecteur, 67
Geoid, 161, 162, 165
Geomechanique, 126
Geometrics Strativisor NX, 24, 27, 36, 43
Geopulse, 111
Geo-resources, 123, 124
Geospace DAS 1A, 24
GI airgun, 52, 53
GLONASS, 156, 157, 175
GPS, 14, 58, 91, 156, 156, 157, 158, 159, 160, 161, 165, 166, 168, 170, 172, 173, 175, 177
grab buckets, 11
gravity, 11, 13, 141, 143, 145
Greystoke current meter, 138
guide base stability, 16, 21
gyro, 19, 102, 104, 141, 143, 186, 188

harmonic distortion, 42
Hookes Law, 1
horizontal accuracy, 6
Hugin, 106
Huntec, 118
hydrographic echosounder, 8, 81, 84
Hyperfix, 14

IBM, 34
IFP, 30, 31, 32, 39
interchannel crossfeed, 32, 41, 62
iso-velocity, 67

jack-up, 8, 9, 15, 16, 17
JAMSTEC, 108

Kalamos, 64
K-check, 145
Kirchoff migration, 71
Klein, 90, 111
Komsonolets, 109
Krohn-Hite filter, 131
Kullenberg gravity corer, 134

Lacoste and Romberg, 141
Lamour frequency, 146
Langevin, 2
leakage, 56, 64, 85, 127, 133
Least Astronomic Tide (LAT), 21, 85, 104
Liberty Bell, 14
Lithology, 21
long range real-time kinematic (LRTK), 155
LRS-100, 58

Macartney Focus, 139
magnetometers, 11, 105, 140, 145, 147, 148, 151, 153
magnetostrictive, 2, 109
manometer, 56
Manta-Ceresia, 108
Mantissa, 24
Marconi, 107
Martin 200, 105, 106
Maxiran, 14
MDS-10, 24
Medwin's formula, 84, 85
Micromax, 40, 67, 133
microprocessor, 31, 58, 65, 90
migration, 6, 71
mini-streamer, 9, 34
mis-ties, 6
mnemonic, 30
Monitor (warship), 14
monitor (warship type), 2
multi-electrode sparker, 9, 81, 93, 123, 124, 125, 126, 130
multiple, 3, 8
multiplexer, 7, 24, 25, 28, 31, 32, 37, 38, 39, 112
mute, 67, 70, 73

NAVSTAR, 14, 155, 161
normal moveout (NMO), 67, 73
North Sea, 9, 10, 11, 18, 21, 44
Nyquist, 42

Ocean Seven, 84
Odom Echotrac, 81

Odyssey, 107
OKPO 6000, 109
opto-isolator, 50
ORE, 111
outcroppings, 8
OYO Geospace, 24, 28, 34, 36, 38, 40, 41, 42, 43, 130
OYO plotter, 34

P-code, 156
Penetrometers, 135, 136
piezocone penetrometer test, 138
piezoelectric, 2
pingers, 8, 9, 11, 13, 14, 15, 81, 104, 109, 111, 112, 131
pipelines, 8, 13
plankton sampling, 11
pock marks, 8, 134
Polaroid, 58
pollution sampling, 11
precision (P) code, 14
profiler, 8, 9, 11, 13, 14, 15, 81, 93, 105, 109, 111, 112, 123, 124, 131, 140
Promax, 40, 67, 133
pseudo-range, 156, 160, 166, 175
PTEROA-150, 109
pulse repetition rate (PRP), 8
P-Waves, 1, 2, 77

quality control supervisor, 21
Quantum DAS 1A, 24
quaternary, 7, 24

Racal, 112, 140, 155, 156
Ramasses, 67
Rayleigh, 2
Rayleigh-Willis, 2
real time kinematic (RTK), 155, 157, 166, 169
refraction, 2, 12, 1, 131, 132
relative amplitudes, 24
remote operated vehicles (ROVs), 11, 13, 14, 97, 137, 138, 169, 186, 190
resolution, 6,8

Safre-Crouzet C10, 84, 85
salt dome, 2
sample skew, 32
sampling, 7, 32, 34, 37
Seaeye Surveyor, 137
seawater sampling, 11
SEG A, 25
SEG B, 25, 32
SEG C, 25

Index

SEG D, 25, 31, 34
SEG Y, 25, 72, 112, 132
seismic interference, 6
seismic pulse, 9
seismic source, 6
Seismic Systems Inc., 52
seismographs, 2
Senkovitch corer, 136
Sercel SN 358 DMX, 24, 28, 30, 31, 37
shotpoint, 7, 12, 21, 28, 34, 60, 77
sidescan sonar, 8, 13, 14, 15, 21, 81, 85, 88, 92, 93, 95, 101, 104, 105, 107, 109, 111, 112, 134, 135, 140, 141, 186
sigma-delta, 7, 24, 28, 31, 34, 36, 37, 38, 40
Simrad, 97, 106, 107
skew, 40
Skyfix, 172
sledge, 12, 13
sleeve guns, 7, 51, 57
Snells Law, 12
SOLAS, 192
sonar swath bathymetry, 13
sonobuoys, 12, 77
spans, 13
spark arrays, 7
sparkers, 7, 9, 43, 44, 45, 46, 47, 48, 50, 51, 52, 58, 60, 115, 119, 121, 123, 126, 130, 131, 197
spheroid, 160, 164
Starfix, 156, 172
statics, 67
Statoil, 106
Strativisor *see* Geometrics Strativisor, 28, 36,
Stealth 3000 ROV, 141
streamers, 6, 7, 13, 18, 19, 28, 32, 37, 38, 60, 62, 64, 65, 66, 67, 77, 78, 79
submersibles, 13
Submetrix ISIS 100, 101
sub-tow, 9
super short baseline (SSBL), 108
survey grid, 8, 10, 15, 16
swath bathymetry, 10, 13, 95, 97, 99, 101, 103, 140, 169, 186
S-Waves, 1
Syledis, 14

Syntron, 62
SYS 100, 141
system noise tests, 38

tailbuoy, 5, 62, 66, 77, 93
tape deck, 26, 29, 31, 38
tape transport, 25, 28, 30, 40, 78
Teledyne, 7, 57, 60, 62
telephone cables, 8
test oscillator, 32
Texas Instruments 960A, 24
Theseus, 107
time variant gain, (TVG), 93, 111
Titanic, 14, 81
trace sequential data, 25
triode, 2
TTS-2, 24, 26, 28, 31, 32, 34, 37, 40, 42, 43
Twinburger, 108

UKOOA, 67, 172, 175
Ultra Short Baseline (USBL), 91
UMEL Sargent, 134
underwater photography, 11
underwater video, 11
UROV7K, 108

Vab Steen grab bucket, 134
Valeport, 84
variable pitch propellers, 19
velocity analysis, 67, 73
velocity data, 12
Veripos, 156, 172
vertical accuracy, 6
vibro-corers, 11
VORAM, 109

water guns, 7, 51, 57
wavelet, 6, 67, 71, 72, 111
wellheads, 11
West Africa, 18
Willis, 3
Wilson's formula, 84
W-test, 172, 176

Zeeman effect, 149
Zenith, 66

For Product Safety Concerns and Information please contact our EU
representative GPSR@taylorandfrancis.com
Taylor & Francis Verlag GmbH, Kaufingerstraße 24, 80331 München, Germany

www.ingramcontent.com/pod-product-compliance
Ingram Content Group UK Ltd.
Pitfield, Milton Keynes, MK11 3LW, UK
UKHW021441080625
459435UK00011B/329